KU-408-841

BRITISH GEOLOGICAL SURVEY

C R BRISTOW,
E C FRESHNEY and I E PENN

Geology of the country around Bournemouth

Memoir for 1:50 000 geological sheet 329
(England and Wales)

CONTRIBUTORS

Palaeontology
D K Graham
R Harland
M Hughes
C J Wood

Stratigraphy
B J Williams

Water supply
R A Monkhouse

LONDON: HMSO 1991

© *NERC copyright 1991*

First published 1991

ISBN 0 11 884377 X

Bibliographical reference

Bristow, C R, Freshney, E C, and Penn, I E. 1991. Geology of the country around Bournemouth. *Memoir of the British Geological Survey*, Sheet 329 (England and Wales).

Authors

C R Bristow, BSc, PhD
E C Freshney, BSc, PhD
British Geological Survey, Exeter

I E Penn, BSc, PhD
British Geological Survey, Keyworth

Contributors
B J Williams, BSc
formerly *British Geological Survey*
R Harland, DSc
British Geological Survey, Keyworth
D K Graham, BA
British Geological Survey
Edinburgh
R A Monkhouse, BA, MSc
British Geological Survey, Wallingford
M J Hughes, MSc and C J Wood, BSc
formerly *British Geological Survey*

Printed in the UK for HMSO

Dd 291130 C10 10/91

Other publications of the Survey dealing with this district and adjoining districts

BOOKS

British Regional Geology
The Hampshire Basin and adjoining areas, 4th edition

Memoirs
Shaftesbury (313), 1923*
Ringwood (314), 1902*
Southampton (315), 1987
Dorchester (328), 1899*
Bournemouth (329), 1917*
Lymington (330), 1915*
Weymouth (342), 1947
Swanage (343), 1947

Mineral Assessment Reports
No. 51 (Bournemouth), 1981
No. 103 (Dorchester and Wareham), 1982

Water Supply

The Water Supply of Hampshire (including the Isle of Wight), published 1910*
Wells and springs of Dorset, published 1916*

1:584 000

Tectonic map of Great Britain and Northern Ireland.

1:100 000
Hydrogeological map of the Chalk and associated minor aquifers of Wessex

1:100 000
South Sheet (Geological)
South Sheet (Quaternary)
South Sheet (Aeromagnetic)

1:250 000
Wight (Solid geology)
Wight (Sea bed sediments)
Wight (Gravity)
Wight (Aeromagnetic)
Portland (Solid geology)
Portland (Sea bed sediments)
Portland (Gravity)
Portland (Aeromagnetic)

Maps

1:50 000 and 1:63 360
Sheet 313 (Shaftesbury) (1923) (revised edition *in press*)
Sheet 314 (Ringwood) (1902)
Sheet 315 (Southampton) (1987)
Sheet 329 (Bournemouth) (1991)
Sheet 330 (Lymington) (1893)
Sheet 342 (Weybridge) (1949)
Sheet 343 (Swanage) 1895)

* Out of print

Geology of the country around Bournemouth

The Bournemouth district in east Dorset lies towards the western end of the Hampshire Basin. The central part of the area is occupied by the ever-growing urban conurbation of Poole–Bournemouth–Christchurch. Beyond the conurbation, the area is one of contrasting scenery which reflects the underlying geology. The area is crossed by the rivers Stour and Avon, which empty into Christchurch Harbour, and the River Frome and Sherford River which flow into Poole Harbour. Both these harbours provide anchorage for the large boating community, both private and commercial. South of Poole Harbour is the Arne Peninsula, a nationally important area of heath and marshland. The Wytch Farm Oilfield, the largest onshore field in western Europe, is situated in the extreme south-west of the district.

The geological sequence is dominated by Tertiary deposits. Chalk crops out over a small area in the north-west, and extensive tracts of alluvium and river terrace deposits obscure much of the solid geology in the central part of the district. Deep boreholes and geophysical data have been interpreted to provide a detailed description of the concealed formations. The stratigraphical account of the Tertiary deposits describes the various formations present and, in particular, provides details of a refined stratigraphy of some of the Tertiary formations.

REFERENCE LIBRARY
THIS COPY IS FOR REFERENCE IN THE LIBRARY ONLY AND IS NOT AVAILABLE FOR HOME READING.

COUNTY LIBRARY
19/20 WESTMINSTER HOUSE
POTTERS WALK
BASINGSTOKE RG21 1LS

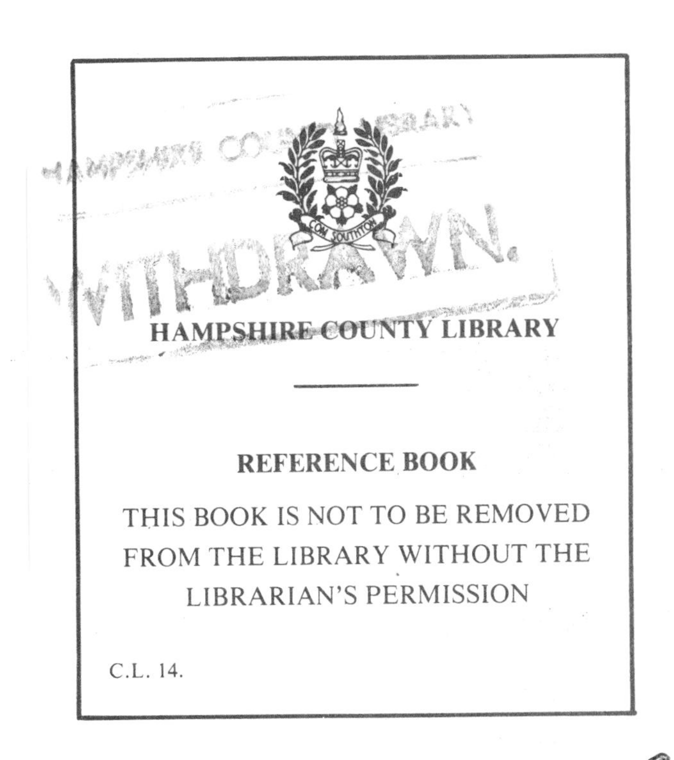

In the tabular geological succession opposite, the columns show, from left to right, system, group, formation and lithological summary, member/bed and lithological summary (with thickness), and generalised thickness of formation. Formations below the Upper Chalk are not exposed in the district, and are encounctered only in boreholes. Thicknesses for formations below the Chalk are those proved in the Wytch Farm, Arne G1 and Shapwick boreholes (see Figure 6)

Cover photograph
Cliff section in Branksome Sand and Boscombe Sand (top of cliff in foreground), Bournemouth. Poole Harbour, with Brownsea Island at the top left of the photograph (*Kitchenham Ltd, Bournemouth*).

Nodding donkeys at Wytch Farm Oilfield.

CONTENTS

FIGURES

TABLES

PLATES

PREFACE

Bournemouth and Poole together represent one of the most rapidly expanding urban areas in the United Kingdom. This memoir describes the geology of the district covered by the Bournemouth 1:50 000 (329) geological sheet, published in 1991. Most of the district is underlain by Tertiary sedimentary deposits. Detailed stratigraphical sequences have been delineated on the map and described in the memoir, providing a sound framework for anticipating and solving geological problems likely to be encountered during the continued, rapid urban expansion. Potential hazards arising from landslips, extensive areas of landfill, solution-collapse hollows developed on Tertiary formations close to the outcrop of Chalk, and former areas of underground clay extraction have been identified. Areas of bulk minerals such as sand, gravel and brickclay are described. Of some economic importance to the areas is the Wytch Farm Oilfield, the largest onshore oilfield in Western Europe. Its structure and stratigraphy is described in detail.

Peter J Cook, DSc
Director

British Geological Survey
Keyworth
Nottingham
NG12 5GG

1 April 1991

ACKNOWLEDGEMENTS

The district was first geologically surveyed on the one-inch scale by H W Bristow and J W Trimmer and the results published on Old Series Sheets 15 and 16 in 1856 and 1855 respectively. The primary six-inch geological survey area was made by C Reid and incorporated in New Series One-inch Geological Sheet 329, published in 1895 in both Solid and combined Solid and Drift editions. The Solid edition was discontinued when the map was colour-printed in 1904. The accompanying memoir, by Reid, was published in 1898 with a second edition, by H J O White, published in 1917. In 1976 the sheet was republished without revision at the 1:50 000 scale. In 1983 the resurvey being reported in this memoir was begun, nearly a century since the district was last geologically mapped. The geological resurvey of the Bournemouth area was made by Dr C R Bristow, Dr E C Freshney and Mr B J Williams between 1983 and 1986 under the direction of Drs R W Gallois and P M Allen, Regional Geologists and supported by the Department of the Environment.

As well as accounts by the field officers, the memoir contains contributions on the macropalaeontology of the Chalk and London Clay by Mr C J Wood, on the macropalaeontology of the Alluvium by Mr D K Graham, on the foraminifera of the Tertiary formations by Mr M J Hughes and on the dinoflagellate cysts by Dr R Harland. Mr S C A Holmes visited many of the chalk pits in the district in the 1930s and made an extensive collection of fossils which he presented to the Survey in 1936. Details of this material are included in this account. The section on hydrogeology has been compiled from reports by Mr R A Monkhouse.

The structure contour maps of the base of the Paleogene and of the base of the Lower Greensand/Gault subcrop, based on seismic data, were prepared by Mr R A Chadwick, Dr S Holloway, Mr G Roberts and Mr A G Hulbert of the Deep Geology Research Group. Dr I E Penn of the same group supplied accounts of the concealed formations, based on the Wytch Farm Oilfield borehole data, and of the hydrocarbon resources; also Figures 3 to 5, showing the structure and correlation of concealed strata. Figure 17 is based on notes and sections measured by Drs R A Edwards and E C Freshney in 1981.

The authors thank Bournemouth, Poole and Wimborne borough councils, and Dorset and Hampshire county councils and their officers for their ready co-operation in providing borehole and other subsurface information from their records. In particular, Bournemouth Borough Council have generously allowed us to use the section of Bournemouth cliffs constructed by Mr R Agar (incorporated in Figure 16), the former Deputy Borough Engineer. E C C Ball Clays Ltd have kindly allowed us to reproduce the gamma-ray log in Figure 11, and to incorporate information from their Rempstone Borehole; in particular, Mr Q G Palmer has provided local detail, and helped with the interpretation of the stratigraphical sequences on the Arne Peninsula. British Petroleum Co. have permitted us to incorporate Tertiary well data from their Bransgore and Hurn boreholes in Figure 12, and their seismic data in Figure 4.

1:10 000-SCALE MAPS

The component 1:10 000-scale National Grid sheets of Geological Sheet 329 are shown on the diagram below, together with the dates of survey and the initials of the surveyors. The surveying officers were C R Bristow, E C Freshney and B J Williams. Uncoloured dyeline copies of the maps are available for purchase from the British Geological Survey, Keyworth, Nottingham NG12 5GG.

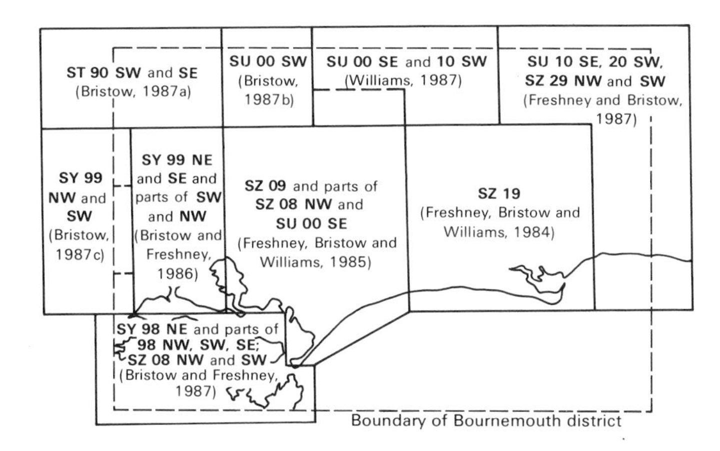

OPEN-FILE REPORTS

Open-file reports, which are available for the whole of the Bournemouth district, are shown on the diagram below, together with the dates of publication and the authors' names. They contain more detail than appears in the memoir. Copies of these reports are available for purchase from the British Geological Survey, Keyworth, Nottingham NG12 5GG and at 30 Pennsylvania Road, Exeter, Devon EX4 6BX.

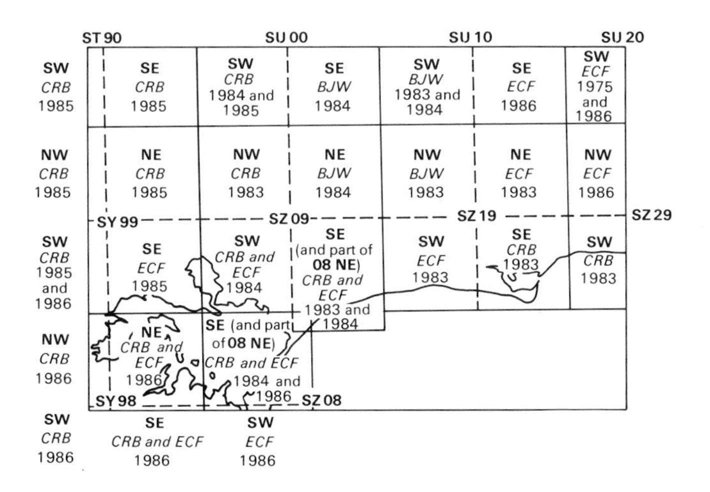

NOTES

The word 'district' used in this memoir means the area included in 1:50 000-scale Geological Sheet 329 (Bournemouth).

Figures in square brackets are National Grid references; places within the Bournemouth district lie within the 100-km squares ST, SU, SY and SZ. The grid letters precede the grid numbers.

The authorship of fossil species is given in the index of fossils.

Numbers preceded by A refer to photographs in the Geological Survey collections.

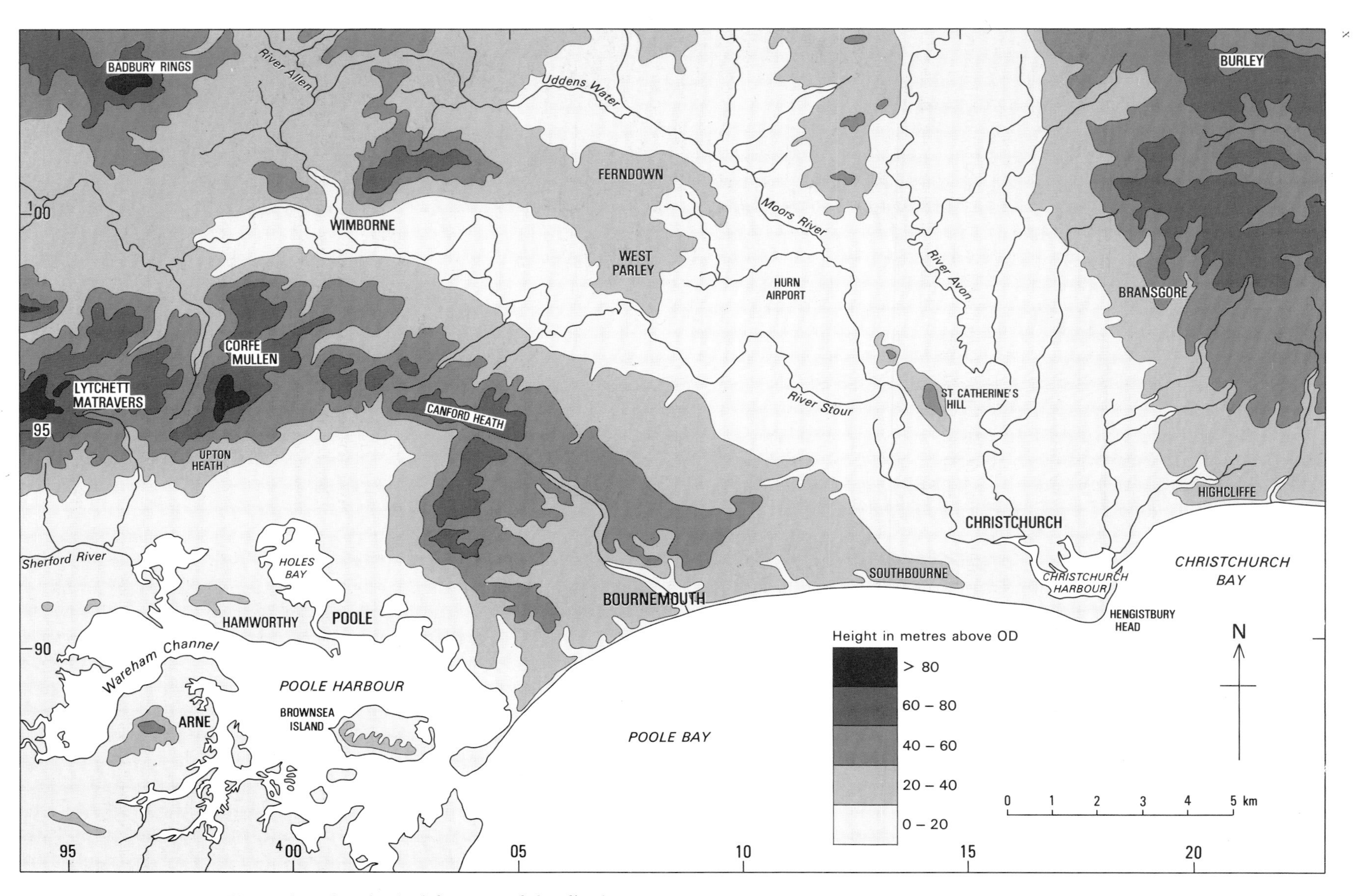

Figure 1 Sketch-map illustrating the physical features of the district.

ONE

Introduction

GEOGRAPHICAL SETTING

The Bournemouth district is one of gentle relief, and nowhere rises above 100 m (Figure 1). The principal drainage is by the River Stour, which flows eastwards and then south-eastwards across the district. It is joined by the southward-flowing River Avon before flowing into Christchurch Bay just east of Christchurch. In the south, the rivers Piddle and Frome unite just west of the district and flow into the sea via the Wareham Channel and Poole Harbour. The watershed between the Stour and the rivers to the south crosses the district in an arc from Lytchett Matravers in the west, where it rises to about 90 m above OD, across Canford Heath at about 65 m above OD to the coast at Boscombe.

The centre of the district is dominated by the conurbation of Poole – Bournemouth – Christchurch, and the town of Wimborne. Outside these areas, the district is rural, with a scattering of small villages set in mixed arable and pasture farmland. In the south-west, much of the land underlain by light Tertiary (Palaeogene) sands remains as heath, although some large areas have been planted with conifers.

The outcrop distribution of solid rock formations is shown in Figure 2. The combination of low relief and unconsolidated sediments ensures that there are few natural exposures inland; the more important of the Palaeogene fossil localities are listed in Table 1.

GEOLOGICAL HISTORY

The oldest known rocks in the district, proved in the Wytch Farm X14 Borehole [SY 9804 8526], are phyllites, which were metamorphosed during the mid to late Devonian. No later Devonian or Carboniferous rocks have been proved in the district. After folding and low-grade metamorphism during the Variscan Orogeny, the Devonian rocks were uplifted and eroded. The oldest postorogenic strata are the red-bed alluvial fan and floodplain deposits of the Wytch Farm Breccias and Aylesbeare Mudstone Group, of presumed Permian age. Further dominantly red floodplain deposits, represented by the Sherwood Sandstone Group, were laid down in the Triassic Period. The succeeding reddish brown, silty mudstones and siltstones of the Mercia Mudstone Group were probably laid down in playa lakes. At the end of this period of red-bed deposition, a major marine transgression occurred in the late Triassic which ushered in a period of marine sedimentation that lasted throughout the Jurassic period.

Jurassic sedimentation in the district took place in shallow seas on a broad continental shelf in a warm climate. Worldwide changes in sea level, combined with local tectonic events, led to a series of transgressions and regressions. Most of the Jurassic sediments of the district are marine clays with subordinate beds of sandstone and shelly oolitic limestone which were deposited in near-shore environments.

At the end of the Jurassic, earth movements caused uplift and erosion, with the removal of most of the Kimmeridge Clay and all of the Portland, Purbeck and Wealden Beds. Another widespread transgression introduced the sands of the Lower Greensand and subsequently the marine clays of the Gault. Shallow-water sedimentation during the deposition of the Upper Greensand was followed by the transgression of the Chalk sea.

Earth movements and subsequent erosion in the late Cretaceous caused some of the Upper Chalk to be removed prior to the deposition of the Tertiary sediments. Little folding occurred, and there is generally no obvious angular discordance between the Cretaceous and Tertiary rocks.

The London Clay is composed of deposits laid down in a restricted marine environment, probably close to the western margin of the London Clay basin. The succeeding Poole Formation consists of an alternating sequence of marine lagoonal clays, interbedded with fluviatile and beach-barrier sands. During the deposition of the succeeding Branksome Sand, fluviatile sedimentation was dominant, with only minor marine incursions. Marine conditions returned when the fine-grained sands and localised beach pebble beds of the Boscombe Sand, the basal formation of the Barton Group, were laid down. Deposition in a shallow marine environment continued with the glauconitic, sandy, shelly Barton Clay. The Barton Clay passes up into the shelly, clayey Chama Sand, which is overlain by the clean, fine-grained Becton Sand, a shoreline facies at the top of the Barton Group. The youngest Palaeogene strata are the freshwater to brackish water lagoonal sands and clays of the Headon Formation.

About half of the district is covered by drift deposits. These are chiefly river terrace sediments, consisting of sand and gravel of potential economic importance. The older river terrace deposits (Thirteenth to Tenth, or Ninth) are related to the 'proto'-Solent river and its tributaries which flowed eastwards across the district, thence across what is now the northern part of the Isle of Wight, to flow into the English Channel some distance to the east of the island (see Reid, 1902, fig.3). The floodplains of the rivers and the estuarine creeks are floored principally by alluvium, although peat is present along the River Allen and, locally, in some of the valleys in the south-west. Extensive areas of blown sand occur along the coast.

PREVIOUS RESEARCH

Little has been published on the geology of the Bournemouth district, except for the coastal cliff sections. The strata in those between Christchurch [Hengistbury] Head and

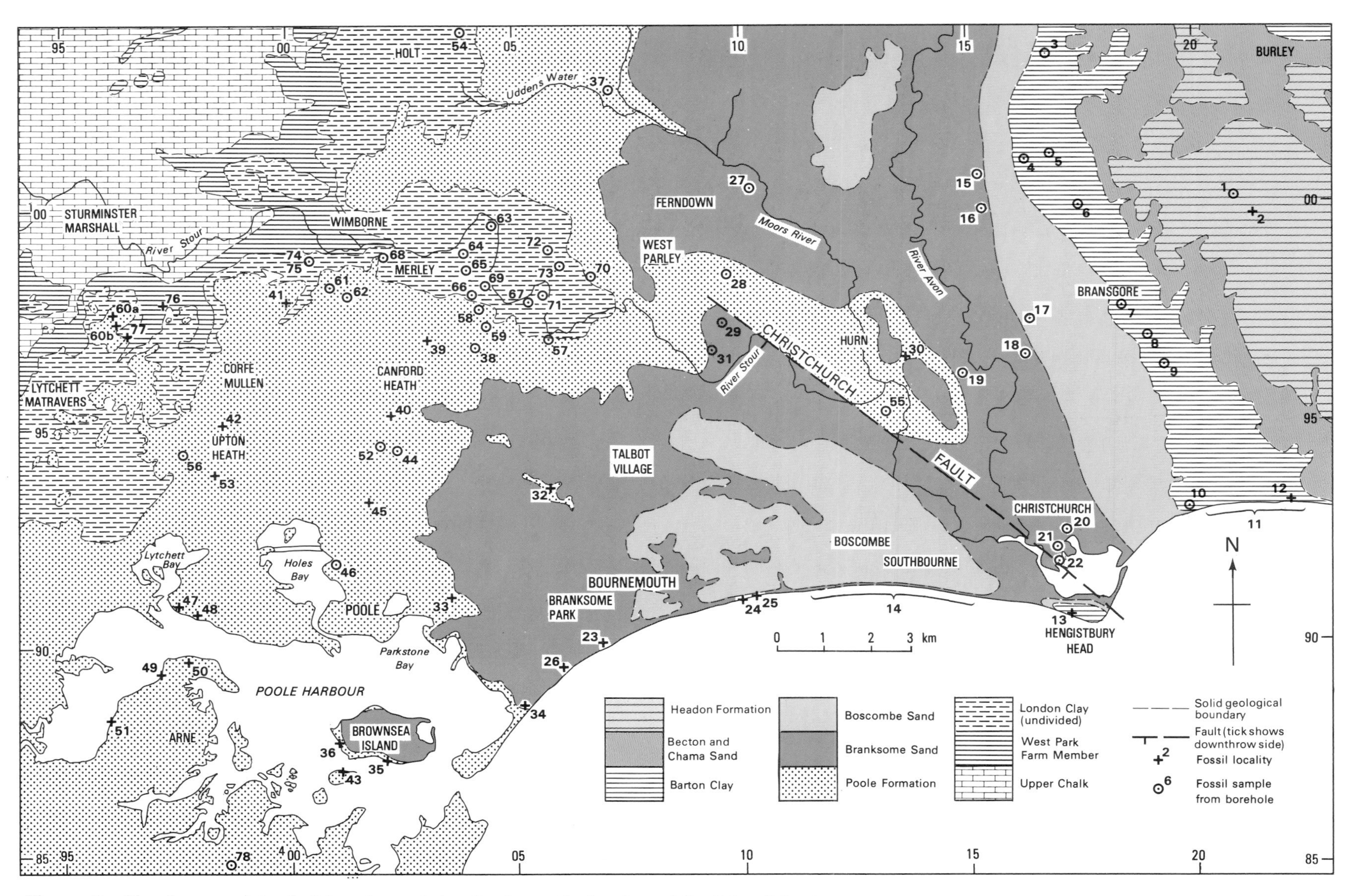

Figure 2 Sketch-map of the Solid geology of the district, with the sites of fossil localities (see Table 1 for list of fossil localities).

Table 1 Palaeogene fossil localities in the district (See Figure 2 for sites of fossil localities).

Locality		Grid reference	Stratal unit	Fossil group	Zone
1	Thorney Hill, Holms	SU 2131 0015	Headon Formation	Dinoflagellates	non-diagnostic
2	Holmsley Inclosure	SZ 217 997	Headon Formation	Dinoflagellates	non-diagnostic
3	Upper Kingston Farm	SU 1682 0330	Barton Clay	Dinoflagellates	late Middle Eocene or younger
4	Broad Heath	SU 1633 0100	Barton Clay	Dinoflagellates	*coleothrypta* or *intricata* Ass. Zone
5	Hain Hill	SU 1696 0104	Barton Clay	Dinoflagellates	non-diagnostic
6	Whistler's Copse	SZ 1752 9975	Barton Clay	Dinoflagellates	*porosum*, or slightly younger
7	Bransgore	SZ 1853 9766	Barton Clay	Dinoflagellates	*porosum* [or *intricata*]
8	Godwinscroft	SZ 1905 9685	Barton Clay	Dinoflagellates	non-diagnostic
9	Allensworth Wood	SZ 1952 9627	Barton Clay	Dinoflagellates	*laticinctum* or younger
10	Christchurch Borehole	SZ 2002 9301	Barton – London Clay	Dinoflagellates	*porosum* to *simile*
11	Barton Cliffs	SZ 200 929 to 230 930	Barton Clay to Headon Fm.	Abundant molluscs some vertebrates	
12	Highcliffe	SZ 219 931	Barton Clay	Dinoflagellates	*draco*
13	Hengistbury Head	SZ 1690 9065 to 1780 9040	Barton Clay – Boscombe Sand	Dinoflagellates	*draco* to *intricata* Ass. Zone
14	Bournemouth cliffs	SZ 1200 9125 to 150 910	Boscombe – Branksome Sand	Dinoflagellates	*coleothrypta (intricata* Ass. Zone)
15	Bisterne Manor	SU 1550 0062	Branksome Sand	Dinoflagellates	Early Eocene or younger
16	Anna Lane	SZ 1550 9964	Branksome Sand	Dinoflagellates	non-diagnostic
17	Clapcott's Farm	SZ 1632 9723	Branksome Sand	Dinoflagellates	non-diagnostic
18	near Winkton	SZ 1637 9644	Branksome Sand	Dinoflagellates	non-diagnostic
19	Dudmoor Farm	SZ 1494 9593	Branksome Sand	Dinoflagellates	non-diagnostic
20	Stanpit 1	SZ 1715 9235	Branksome Sand	Dinoflagellates	Early Eocene
21	Stanpit 2	SZ 1697 9212	Branksome Sand	Dinoflagellates	Early Eocene
22	Stanpit 3	SZ 1700 9168	Branksome Sand	Dinoflagellates	Early Eocene
23	Branksome Dene	SZ 069 899	Branksome Sand	Dinoflagellates	*coleothrypta* or younger
24	Eastcliff	SZ 1000 9100	Branksome Sand	Dinoflagellates	*coleothrypta (intricata* Ass. Zone)
25	Eastcliff	SZ 1025 9106	Branksome Sand	Dinoflagellates	*coleothrypta (intricata* Ass. Zone)
26	Canford Cliff	SZ 0606 8931	Branksome Sand	Dinoflagellates	*coleothrypta* or younger
27	Parley Court No. 2	SU 1020 0015	Parkstone Clay	Dinoflagellates	*coleothrypta* or younger
28	Barnes Farm	SZ 0965 9830	Parkstone Clay	Dinoflagellates	non-diagnostic
29	Parley Court	SZ 0962 9715	Parkstone Clay?	Dinoflagellates	non-diagnostic
30	Blackwater Hall Bridge	SZ 1350 9641	Parkstone Clay	Dinoflagellates	Early Eocene
31	Parley Court No. 1	SZ 0928 9649	Parkstone Clay	Dinoglagellates	*coleothrypta* or younger
32	Rossmore	SZ 0582 9355	Parkstone Clay	Dinoglagellates	*coleothrypta* or younger
33	Parkstone	SZ 0388 9104	Parkstone Clay	Dinoflagellates	non-diagnostic
34	'Sandbanks'	SZ 051 884	Parkstone Clay	Macroflora	non-diagnostic
35	Brownsea Island	SZ 021 875	Parkstone Clay	Dinoflagellates	*coleothrypta* or younger
36	Brownsea Island	SZ 0100 8797	Parkstone Clay	Dinoflagellates	*coleothrypta* or younger
37	Dolman's Farm	SU 0713 0256	Poole Formation undifferentiated	Dinoflagellates	non-diagnostic
38	Canford Magna No. 8	SZ 0429 9668	Broadstone Clay	Dinoflagellates	Early Eocene
39	Canford Heath	SZ 0305 9670	Broadstone Clay	Dinoflagellates	non-diagnostic
40	Canford Heath (S)	SZ 0208 9508	Broadstone Clay	Dinoflagellates	non-diagnostic
41	Ashington	SZ 0004 9787	Broadstone Clay	Dinoflagellates	*coleothrypta* or younger
42	Beacon Hill	SY 9834 9523	Broadstone Clay	Dinoflagellates	*coleothrypta* or younger
43	Furzey Island	SZ 0115 8720	?Broadstone Clay	Dinoflagellates	non-diagnostic
44	Canford Heath	SZ 0248 9430	Oakdale Clay	Dinoflagellates	*coleothrypta* or younger
45	Oakdale	SZ 0246 9453	Oakdale Clay	Dinoflagellates	*coleothrypta* or younger
46	Sterte Borehole	SZ 0095 9200	Oakdale Clay	Dinoflagellates	*coleothrypta* or younger
47	Rockley Sands	SY 9737 9090 SY 9739 9108	Oakdale Clay	Dinoflagellates	*coleothrypta* or younger
48	Lake	SY 979 907	Oakdale Clay	Macroflora	non-diagnostic
49	Arne	SY '970 844'	Oakdale Clay	Macroflora	non-diagnostic
50	Arne Clay Pit	SY 976 896	Oakdale Clay	freshwater alga and non-marine dinoflagellates	non-diagnostic
51	Arne Heath	SY 9606 8846	Oakdale Clay	Dinoflagellates	*coleothrypta* or younger
52	Marley Tile Pit	SZ 0195 9438	Oakdale – Creekmoor Clay	Dinoflagellates	*simile* or younger
53	Upton Heath	SY 9809 9400	Creekmoor Clay	Dinoflagellates	non-diagnostic
54	Bowers Farm	SU 0362 0408	London Clay	Dinoflagellates, foraminifera	Division B or C of King

Table 1 continued

Locality	Grid reference	Stratal unit	Fossil group	Zone
55 Holdenhurst Borehole	SZ 1320 9519	London Clay	Dinoflagellates	*simile* or younger
56 Beacon Hill Borehole	SY 9761 9446	London Clay	Dinoflagellates	*varielongitudum-hyperacantha*
			Foraminifera	Division B of King (1981) and older
57 Canford Magna No. 7	SZ 0578 9694	London Clay	Dinoflagellates	*?varielongitudum* or younger
58 Canford Magna No. 4	SZ 0439 9769	London Clay	Dinoflagellates	*simile* or younger
			Foraminifera	non-diagnostic
59 Canford Magna No. 5	SZ 0443 9729	London Clay	Dinoflagellates	*simile* or younger
60a Henbury Pit	SZ 9631 9732	London Clay (top)	Dinoflagellates	*simile* or younger
60b Henbury Pit	SZ 9613 9766	London Clay (top)	Dinoflagellates	*simile* or younger
61 Higher Merley No. 1	SZ 0139 9792	London Clay (top)	Dinoflagellates	*meckelfeldensis* or younger
62 Higher Merley No. 2	SZ 0098 9823	London Clay (top)	Dinoflagellates	*meckelfeldensis* or younger
63 Knighton	SZ 0447 9965	London Clay	Dinoflagellates	*meckelfeldensis* or younger
			Foraminifera	Divisions B and C of King (1981)
64 Canford Magna No. 1	SZ 0400 9908	London Clay	Dinoflagellates	Late Palaeocene/Early Eocene
65 Canford Magna No. 2	SZ 0406 9859	London Clay	Dinoflagellates	Early Eocene
66 Canford Magna No. 3	SZ 0425 9797	London Clay	Dinoflagellates	Early Eocene,
			Foraminifera	non-diagnostic
67 Canford Magna No. 6	SZ 0535 9789	London Clay	Dinoflagellates	Late Palaeocene/Early Eocene
			Foraminifera	c. base of Division B of King (1981)
68 Cruxton Farm	SZ 0207 9870	London Clay	Dinoflagellates	Middle Eocene, non-diagnostic
			Foraminifera	
69 Knighton Farm	SZ 0445 9821	London Clay	Dinoflagellates	Early Eocene
70 Canford Magna No. 9	SZ 0670 9802	London Clay	Dinoflagellates	*hyperacantha* or nearshore environment of deposition
71 Canford Magna No. 10	SZ 0511 9952	London Clay	Dinoflagellates	*hyperacantha* or nearshore environment of deposition
72 Canford Magna No. 11	SZ 0593 9882	London Clay	Dinoflagellates	*hyperacantha* or nearshore environment of deposition
73 Canford Magna No. 12	SZ 0501 9860	London Clay	Dinoflagellates	*hypercantha* or nearshore environment of deposition
74 Merley No. 1	SZ 0055 9870	London Clay (base)	Dinoflagellates	*hyperacantha* or nearshore environment of deposition
75 Merley No. 2	SZ 0050 9887	London Clay (base)	Dinoflagellates	*hyperacantha* or nearshore environment of deposition
76 Knoll Manor	SY 9742 9770	London Clay (base)	Dinoflagellates	*hyperacantha* or nearshore environment of deposition
			Molluscs	?Division A of King (1981)
77 Henbury Plantation	SY 9645 9703	London Clay (base)	Molluscs	?Division A of King (1981)
78 Rempstone	SY 9858 8507	London Clay	Dinoflagellates	*meckelfeldensis-varielongtitudum*
			Foraminifera	Division A to B

Studland Bay were referred to by Lyell (1827) as the 'Plastic Clay Formation'. Later, Prestwich (1849) correctly identified Barton Clay at Hengistbury Head, although this was disputed by Gardner (1879). Gardner (1879; 1882) was the first author to pay particular attention to the cliff sections between Poole Harbour and Hengistbury Head. He reviewed the earlier work in the area, particularly (1882) the research into the flora. Gardner named the strata exposed in the cliff section as the Bournemouth Beds, and divided them into the following ascending sequence: Lower or Freshwater Series, Upper (or Bournemouth) Marine Series (or Beds), Boscombe Sands, Hengistbury Head Beds and Highcliff [sic] Sands (Figure 9). He equated the whole sequence with the Bagshot Beds of the London Basin. It is unfortunate that Gardner (1879) appears to have misread Fisher's (1862) account of the Barton Beds, in which the term High Cliff Sands (later Highcliff Sands) of Wright (1851) was used for sandy strata within the Barton Clay, and used the term Highcliff Sands for strata at Highcliffe below the Barton Clay.

During the original 6-inch mapping of the district (1892 to 1895), Reid divided the Palaeogene sequence into the following units, in ascending order: Reading Beds, London Clay, Bagshot Beds, Bracklesham Beds, Barton Clay, Barton Sands and Headon Beds.

Ord (1914) and White (1917) followed the same basic classifications as Gardner, but differed in their interpretation as to which units were grouped with the Bagshot Beds and which with the Bracklesham Beds. Burton (1933) divided the Barton Clay in the type area into faunal and lithological divisions numbered, in ascending sequence, A1 to A3 and B to F respectively. In 1942, Curry found *Nummulites prestwichianus* near the top of the Hengistbury Beds, i.e. in strata below Gardner's 'Highcliff Sands', thus confirming Prestwich's original correlation of the strata at Hengistbury Head with the Barton Clay of the type locality, and also demonstrating that the so-called 'Highcliff Sands' at Hengistbury are younger than the Highcliff Sands at Highcliffe, which underlie the Barton Clay.

Costa et al. (1976) obtained dinoflagellate cysts from the cliff sections between Bournemouth and Hengistbury Head, and established a correlation between the Bournemouth Marine Beds and the youngest Bracklesham Beds of the Isle of Wight.

Curry et al. (1978) introduced the terms Poole Formation for the pipeclay-bearing strata west of Poole Head, and Bournemouth Formation for the combined Bournemouth Freshwater Beds, Bournemouth Marine Beds and Boscombe Sands. The Poole Formation included a sequence of sands and clays above the London Clay and beneath the Bournemouth Freshwater Beds (the Branksome Sand of Freshney, Bristow and Williams, 1985). The top and bottom of the Poole Formation were, therefore, well defined, although local sequences within the formation were as yet poorly described.

Plint (1983b) referred to the strata between the London Clay and the Barton Formation as the Bracklesham Formation, in which he recognised five sedimentary cycles. Each of the cycles, which are numbered T1 to T5 in ascending sequence, commences with a marine transgression.

The discovery of the Wytch Farm Oilfield in 1973 led to much exploratory drilling and the elucidation of the subsurface stratigraphy and structure (Colter and Havard, 1981).

One of the earliest studies of the Quaternary deposits was by Reid (1902; in White, 1917) who recognised an older drainage system which he named the 'proto'-Solent. Papers by Bury (1933), Green (1946; 1947) and Calkin and Green (1949) were important because they divided the terrace deposits into a number of named levels, in some of which flint implements occurred. Sealy (1955) and Clarke (1981) made detailed subdivisions of the terrace suite, incorporating the results of pebble counts and, in the case of the latter author, grain-size analyses of the deposits. Each author numbered or named the terraces differently from the previous workers; a correlation of these terrace nomenclatures with that adopted during the resurvey is given in Table 2.

Remapping of the Bournemouth district at the 1:10 000 scale began in 1983 as a Department of the Environment contract in the east, and moved progressively westwards. The stratigraphical sequence presented in this account is the culmination of that work (Figures 8 and 9; Freshney et al., 1984, 1985; Bristow and Freshney, 1986a). CRB, ECF

TWO

Concealed geology

INTRODUCTION

This review of the concealed geology of the district is based upon seismic reflection data, geophysical log traces and cutting and core samples from the twenty or so boreholes (Figure 3) which were drilled as part of the exploitation of the Wytch Farm Oilfield in the south-west of the district. Details of only one deep borehole outside the oilfield area, i.e. Shapwick No. 1 [ST 9428 0134], have been released.

Whittaker et al. (1985) showed that post-Carboniferous formations in the United Kingdom can be described and diagnosed by means of characteristic geophysical log signatures, choosing gamma-ray and sonic log traces to epitomise these. In this account, the geophysical log signature of the various lithostratigraphical units refers to this use of the gamma-ray and/or sonic log traces.

Regional knowledge (Whittaker, 1985) shows that the concealed upper crustal geology comprises a ?Permian to Tertiary 'cover' sequence unconformably overlying a deformed Variscan basement. A single deep crustal seismic reflection profile from the western boundary of the district shows the basement to comprise a typical southern British, three-layered crustal section (Figure 4). The lowest seismic zone (zone 3, lower and ?middle crust) shows subhorizontal, high-amplitude, but discontinuous reflection events, between a poorly defined Moho which may be at 10.8 seconds(s) two-way travel-time (TWTT) or even deeper, and a well-defined top at 5.6 s TWTT. Regional studies suggest that the upper part of this zone probably comprises a metamorphic, mid to lower crustal basement, but the lower parts are of unknown composition and structure, though of ductile fabric. The succeeding seismic zone (zone 2, upper crust) between 1.2 s TWTT and 5.6 s TWTT is characteristically featureless and is the seismic response of the Variscan Foldbelt. It is crossed by steeply dipping reflection events commonly interpreted as thrusts. The top of this zone was penetrated in Wytch Farm X14 Borehole [SY 9804 8526], where some 27 m of pale grey to green phyllites with ptygmatic quartz veins have yielded radiometric ages of 337 ± 5 to 357 ± 5 Ma, indicating phases of mid to late Devonian metamorphism (Colter and Havard, 1981, p.501). The uppermost seismic zone (zone 1) consists of subhorizontal, high-amplitude events representing the 'cover' sequence.

Despite the later mid-Cenozoic deformation, the deformed upper surface of the basement shows the major structural features which affected the nature and distribution of the 'cover' sequence in the Bournemouth district and surrounding areas (Figure 5). A series of east–west and

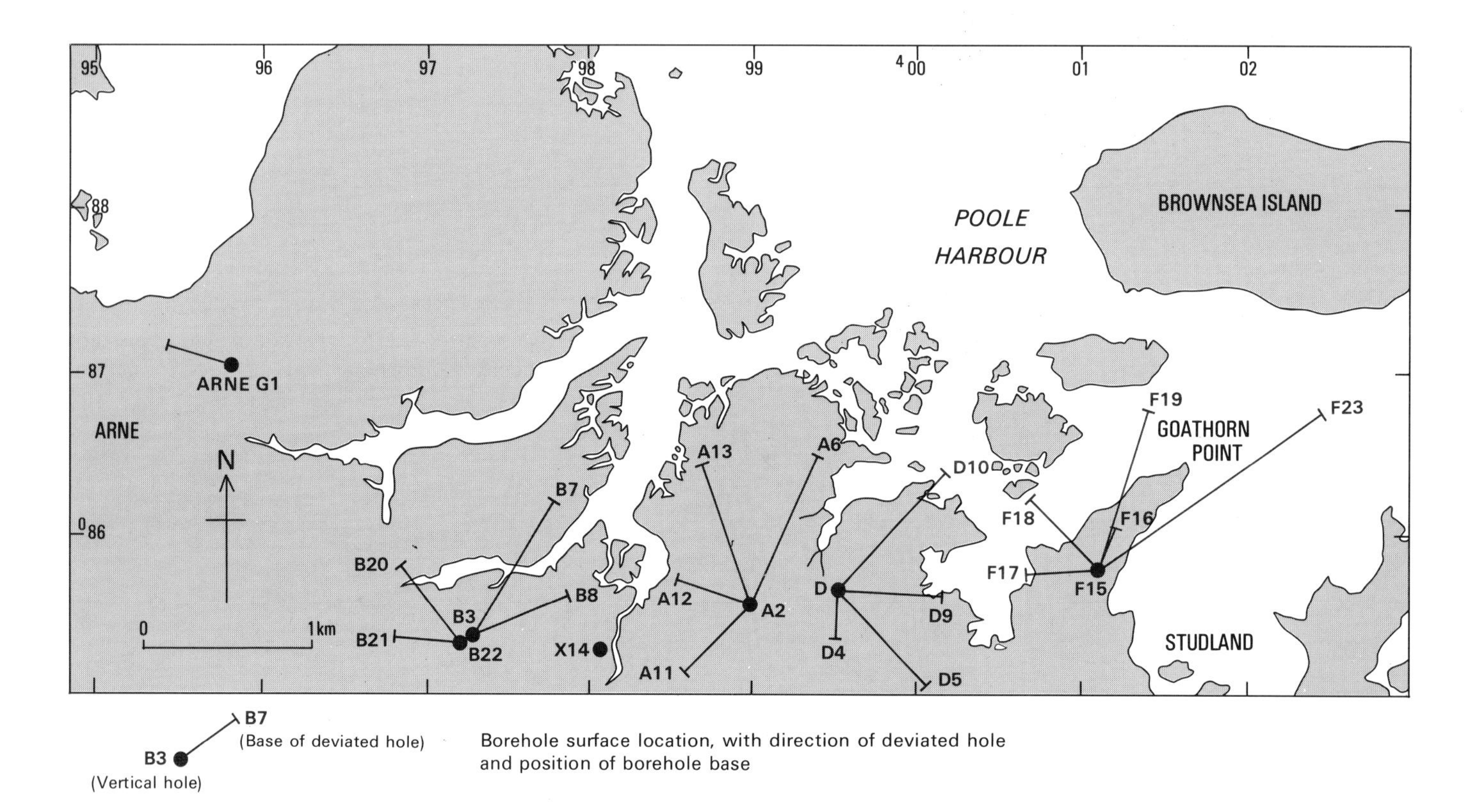

Figure 3 Location of nonconfidential boreholes in the Wytch Farm Oilfield.

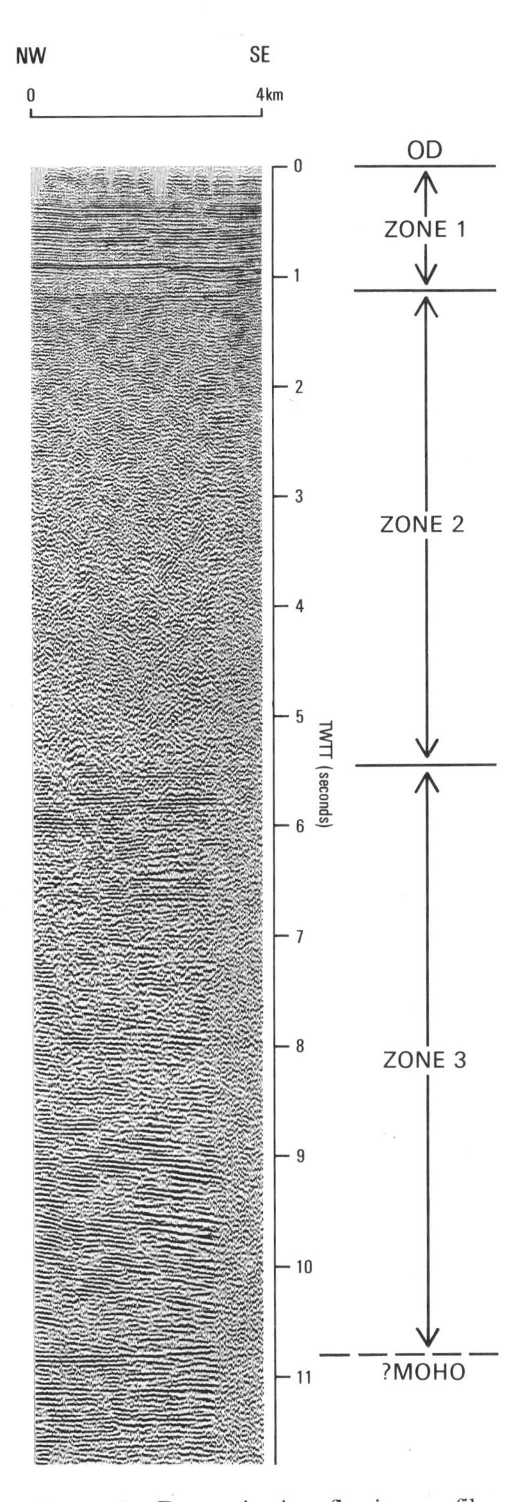

Figure 4 Deep seismic reflection profile, unmigrated, equalised reduced-scale section on western margin of the district.

WNW–ESE-trending faults divides the region into a series of northerly tilted fault blocks, such that a fault-bounded trough, the Winterborne Kingston Trough, is flanked to the north by an area of thinner 'cover', the Cranborne–Fordingbridge High, and to the south by a less well-developed South Dorset High (Penn et al., 1980; Penn, 1982; Figure 5); these major features together comprise the Dorset Basin of Whittaker (1985). The tilted fault-blocks are laterally discontinuous and each plunges westwards; thus the Winterborne Kingston Trough is rather weakly expressed in the district (Figure 5).

?PERMIAN

The strata, comprising the Wytch Farm Breccias below and the Aylesbeare Mudstone Group above, rest with angular unconformity on the Variscan basement. On the general lithostratigraphical resemblance of the Wytch Farm sequence to the now proven Permian strata which crop out in south-west England (Warrington and Scrivener, 1988), the beds have been assigned provisionally to the Permian System (see Smith et al., 1974; Warrington et al., 1980). The sequence reaches a thickness of 925 m in the Wytch Farm area, but thins northwards to zero across the district. There is some indication that the 'feather-edge' is controlled by the Cranborne Fault (Whittaker, 1985, map 4). These ?Permian sediments form three major fining-upward cycles (see below) and were deposited as alluvial fan and floodplain deposits in a continental environment, where red-bed sedimentation was initiated in localised fault-bounded basins (Henson, 1970, 1972; Smith et al., 1974).

Wytch Farm Breccias (48 m)

The base of the ?Permian succession is marked by breccias, the Wytch Farm Breccias of Colter and Havard (1981), composed of large angular fragments (2 to 80 mm across) of grey to green phyllite and quartz pebbles. The formation rests unconformably on the Variscan basement in the Wytch Farm X14 Borehole, where it is some 48 m thick. The breccias in the lowest 19 m are matrix-free, but the clasts in the remainder are set in a matrix of very fine-grained sandstone and silty mudstone, which increases in proportion as the breccias pass upwards into a fine- to coarse-grained, poorly sorted sandstone. The geophysical log signature has a mildly serrated profile (Figure 6), probably reflecting the poorly bedded nature of the breccias. Gamma-ray values are higher than might be expected, and are thought to be due to the high proportion of mudstone fragments. The sonic log of these and the succeeeding ?Permian strata is featureless and registers high velocity values.

Aylesbeare Mudstone Group

Four lithological subdivisions of this group, which ranges from 0 to 865 m in thickness, were recognised by Colter and Havard (1981) in the Wytch Farm area. They are here described in upward succession as follows (Figure 6):

Unit 1 (84 m thick) consists of brick-red, blocky to subfissile mudstones which are sporadically silty and sandy, particularly towards the base. The gamma-ray profile is mildly serrated.

The lower part of Unit 2 (239 m in total thickness) comprises a sequence of fining-upward conglomerates containing pebbles of red to brown mudstone and quartz set in a fine- to coarse-grained sandstone matrix, interbedded with sandstones and siltstones; the upper part consists of very poorly sorted, fine- to coarse-grained sandstones containing sub-

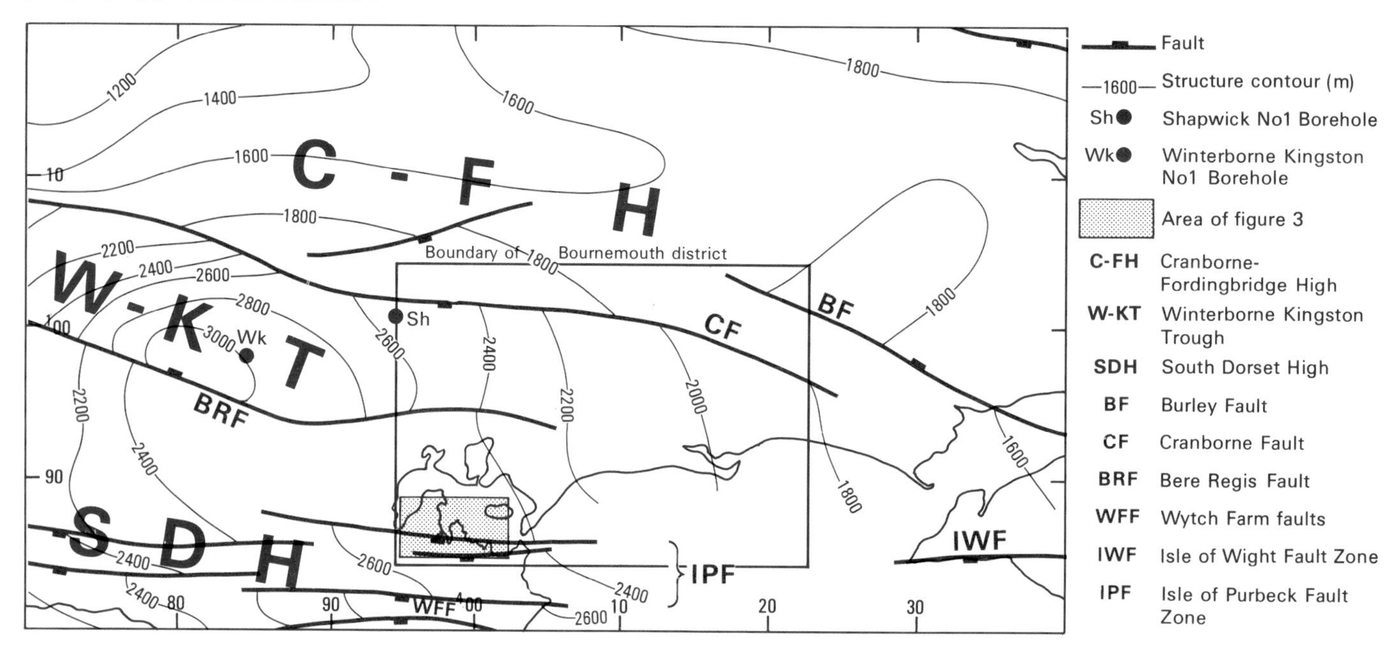

Figure 5 Structural context of the district and surrounding areas showing depth in metres below OD of top of the Variscan Basement, principal Mesozoic structural units and location of boreholes metioned in text.

angular to angular grains, locally with an anhydritic silty mudstone matrix, interbedded with red to brown, sandy, blocky, subfissile mudstones. The interbedded nature of the sequence is reflected in the strongly serrated gamma-ray curve, which also shows decreasing-upward values concomitant with the fining-upward character of the lower part of the sequence.

Unit 3 (151 m thick) is composed mainly of interbedded red to brown, subfissile mudstones and red to purple siltstones, with minor fine- to medium-grained, slightly dolomitic sandstones which contain subangular to subrounded grains and sporadic laminae containing quartz pebbles and mudstone clasts. Unit 3 is a continuation of the upward-fining trend of Unit 2, and this is reflected in its more finely serrated gamma-ray curve and higher gamma-ray values.

Unit 4 (393 m thick) comprises predominantly red to brown, blocky mudstones interbedded with red to purple, finely micaceous siltstones and fine- to medium-grained sandstones with subangular to subrounded grains. The gamma-ray profile is serrated in the lower part of the unit where it characterises small-scale, mainly fining-upward sedimentary cycles of sandstone, siltstone and mudstone.

The Wytch Farm Breccias and the Aylesbeare Mudstone Group together form three major fining-upward cycles, namely Wytch Farm Breccias and Unit 1, Units 2 and 3, and Unit 4.

TRIASSIC

Sherwood Sandstone Group

The Sherwood Sandstone Group is the lower and more productive of the two oil-bearing sandstone reservoirs in the Wytch Farm Oilfield. The sandstone lies with angular unconformity on the Aylesbeare Mudstone Group (Colter and Havard, 1981), and passes upwards into the Mercia Mudstone Group. The Budleigh Salterton Pebble Beds of the Devon coast to the west are absent, whether by overstep or by lateral passage into the sandstone is at present uncertain (Holloway et al., 1989).

The group thins northward across the district from a maximum of 175 m to about 100 m. It thickens towards a basin depocentre located some 50 km to the west, in the Dorchester–Beaminster area [SY 5052 9544] (Allen and Holloway, 1984; Milodowski et al., 1986). The succession is comparable to that proved in Winterborne Kingston Borehole [SY 8470 9796], to the west of the district. There, Lott and Strong (1982) showed that the cored sequence comprises fining-upward, fluvial clastic cycles about 3 m thick, nested within larger cycles, each about 30 m thick. The Sherwood Sandstone Group is thought to have formed as floodplain deposits, the small-scale cycles of which comprise intraformational conglomerates (representing residual, bed-load deposits within river channels), overlain by sandstones (formed as accretionary point bars), and then by siltstones and mudstones (formed as temporary lakes and back levee deposits which were occasionally exposed). Allen and Holloway (1984) suggest that the floodplain may have developed distally from proximal fan deposits, which were banked against the growing Cranborne Fault to the north-east.

Most of the sands are currently provisionally assigned to the Anisian Stage; the uppermost sands are referred to a horizon within the Ladinian Stage (Warrington et al., 1980).

In this district, the group consists predominantly of pale grey to dark grey and red to brown, fine- to very coarse-grained, subangular to rounded, poorly to well-sorted, arkosic sandstones which are variably cemented with calcareous, hematitic or argillaceous material; calcretes occur locally. Anhydrite occurs as an interstitial poikilotopic cement, particularly in the hydrocarbon-bearing sandstones where its corrosion and dissolution gives rise to significant secondary porosity. The sandstones are interbedded with red to brown, massive, soft to hard, micaceous, locally calcareous siltstones, and red to brown, subfissile, micaceous, slightly calcareous mudstones. In the lower half of the group, the sandstones are medium to coarse grained and moderately sorted, and have lower porosity; they are interbedded with thicker mudstones than in the lower part. In the upper half, the sandstones are fine to coarse grained and well sorted, and have higher porosity; there are fewer interbedded siltstones and mudstones. At the top, the sandstones are fine to medium grained and moderately sorted, with low porosity and a high proportion of siltstones and mudstones.

The bedded nature of the sequence is reflected in the deeply serrated gamma-ray curve and the moderately serrated sonic log trace (Figure 6). There is an overall upward increase in gamma-ray values, which corresponds to a general decrease in the grain- size of the sandstones and an upward increase in mudstone content.

Mercia Mudstone Group

The red mudstones of the Mercia Mudstone Group, which is 200 to 600 m thick, can be divided into six persistent lithostratigraphical units (Lott et al., 1982; Whittaker et al., 1985; A to F in Figure 6). The group thickens north-westwards into the Winterborne Kingston Trough where it is over 700 m thick and contains thick halite deposits in Unit C. It thins north-eastwards across the Cranborne Fault, but, because of the north-westerly plunge of the trough, its axial thickness, where it crosses the northern sector of the district, is never more than 500 m, and it decreases continuously east-south-eastwards (Whittaker, 1985, map 8). The mudstones are thought to have been laid down under hot, arid continental conditions in playas or sabkhas.

Within the Wessex Basin, palynomorph assemblages indicate that Units C and D fall within the Carnian Stage, and Unit F yields evidence of a Rhaetian age (Warrington et al., 1980). Thus, it is likely that Units A and B are of Ladinian to Carnian age, and Unit E is of Norian age.

The Mercia Mudstone Group is composed dominantly of reddish brown, firm to soft, more or less silty and sandy mudstones which are variably calcareous or dolomitic. There are minor interbeds of reddish brown, subfissile, calcareous siltstone; colourless, well-rounded, fine- to medium-grained sandstone; white, hard to soft anhydrite; and pale brown, microcrystalline limestone. The dominance of the mudstones produces a characteristically finely serrated geophysical log signature. In the Wytch Farm area, the divisions have the following character from the base upwards:

Unit A is composed of typical reddish brown mudstones, and includes silty and medium- to coarse-grained sandy lenses, particularly in its lowest parts; it is locally difficult to distinguish from Unit B which consists of similar mudstones lacking sandy lenses. The mudstones of Unit B pass up into those of Unit C, which are silty and anhydritic and contain interbedded siltstones.

Unit D has persistent dolomitic and anhydritic sandstone beds at its base, which are thought to be correlatives of the Weston Mouth Sandstone Member of the Devon coast (Warrington et al., 1980). These beds give rise to prominent low gamma-ray and high sonic velocity spikes in the geophysical log profile. They pass upwards, with increasing clay content, into the mudstones of Unit E, the gamma-ray log of which exhibits a finely serrated log curve with high gamma-ray and low sonic velocity values. The signature is characteristically 'waisted', since the mudstone passes gradually upwards into Unit F which comprises alternating grey to green, variably calcareous siltstones and mudstones, giving rise to a highly serrated log profile of lower gamma-ray values and sonic velocity. This unit constitutes the classical 'Tea Green Marl and Grey Marl' sequence of the Blue Anchor Formation in south-western England.

Penarth Group

The Penarth Group is widespread in the district and comprises the Westbury Formation and the overlying Lilstock Formation, the latter consisting of the lower Cotham Member and the upper Langport Member. The group ranges in thickness from about 21 to 25 m. These Rhaetian beds were probably deposited in marginal marine and lagoonal conditions, marking the close of the preceding continental, red-bed sedimentation.

Thin, pale grey sandstone laminae and thin, argillaceous limestone beds, proved in well cuttings, indicate the presence of the Westbury Formation in boreholes at Wytch Farm. The characteristic soft, black shale lithology has not been recorded, but is inferred from the geophysical log trace which is like that of cored boreholes elsewhere (Figure 6). The Cotham Member, consisting of grey to green, calcareous mudstones and grey siltstones, shows the typical overall upward decrease in gamma-ray and sonic velocity characteristic of the beds, but the Langport Member, the most prominent of the Group, consisting of 13 to 14 m of white to pale grey, hard, argillaceous, porcellanous limestone, gives rise to a poorly serrated log profile with low gamma-ray values and correspondingly high sonic velocity. It is a good seismic reflector and forms an excellent marker in subsurface exploration.

LOWER JURASSIC

Lias Group

The Lower, Middle and Upper Lias may be traced throughout the subsurface of the district, although the Middle/Upper Lias boundary is not lithologically well defined, and difficulty is encountered in locating it in boreholes drilled by open-hole methods.

The Lias is generally just over 300 m thick, but it thickens north-westward into and along the axis of the Winterborne Kingston Trough up to a maximum of 450 m, and thins to the north-east across the Cranborne Fault to less than 200 m.

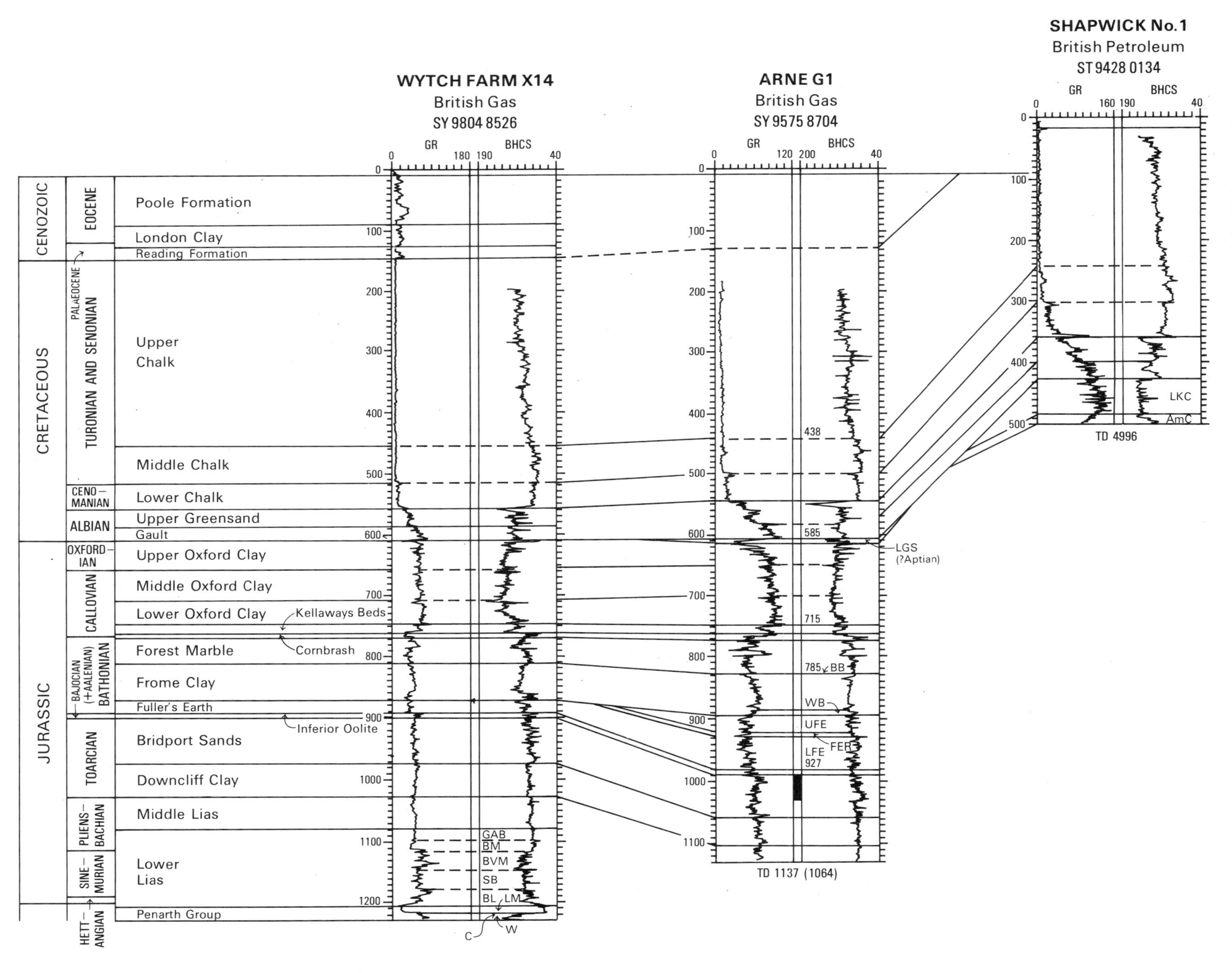
WYTCH FARM X14
British Gas
SY 9804 8526
ARNE G1
British Gas
SY 9575 8704
SHAPWICK No.1
British Petroleum
ST 9428 0134
GR
BHCS
LKC
AmC
TD 4996
LGS
(?Aptian)
438
585
715
785
BB
WB
UFE
FER
LFE
927
TD 1137 (1064)
GAB
BM
BVM
SB
BL
LM
W
C
CENOZOIC
CRETACEOUS
JURASSIC
EOCENE
PALAEOCENE
TURONIAN AND SENONIAN
CENO-MANIAN
ALBIAN
OXFORD-IAN
CALLOVIAN
BAJOCIAN (+AALENIAN) BATHONIAN
TOARCIAN
PLIENS-BACHIAN
SINE-MURIAN
HETT-ANGIAN
Poole Formation
London Clay
Reading Formation
Upper Chalk
Middle Chalk
Lower Chalk
Upper Greensand
Gault
Upper Oxford Clay
Middle Oxford Clay
Lower Oxford Clay
Kellaways Beds
Cornbrash
Forest Marble
Frome Clay
Fuller's Earth
Inferior Oolite
Bridport Sands
Downcliff Clay
Middle Lias
Lower Lias
Penarth Group

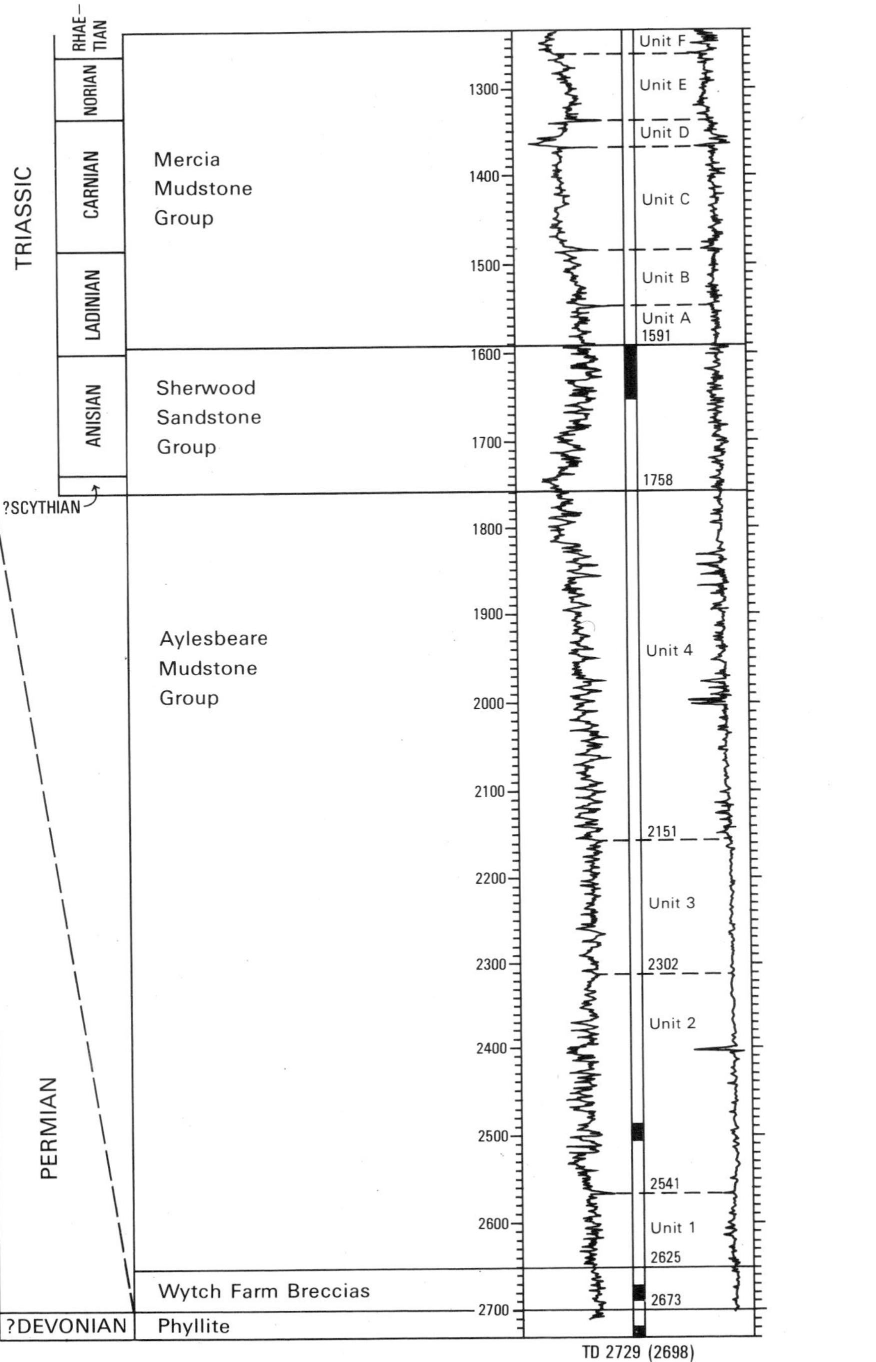

LGS	Lower Greensand
LKC	Lower Kimmeridge Clay
AmC	'Ampthill Clay equivalent'
BB	Boueti Bed
WB	Wattonensis Beds
UFE	Upper Fuller's Earth
FER	Fuller's Earth Rock
LFE	Lower Fuller's Earth
GAB	Green Ammonite Beds
BM	Belemite Marl
BVM	Black Vein Marl
SB	Shales with Beef
BL	Blue Lias
LM	Langport Member
C	Cotham Member
W	Westbury Formation
	Casing shoe
	Core
GR	Gamma-ray log (API units)
BHCS	Borehole compensated sonic log (microseconds/foot)
TD 2729 (2698)	Terminal drilled depth (true vertical depth on right hand side) NB True vertical depths are given within borehole column but scale indicates drilled depths in deviated borehole

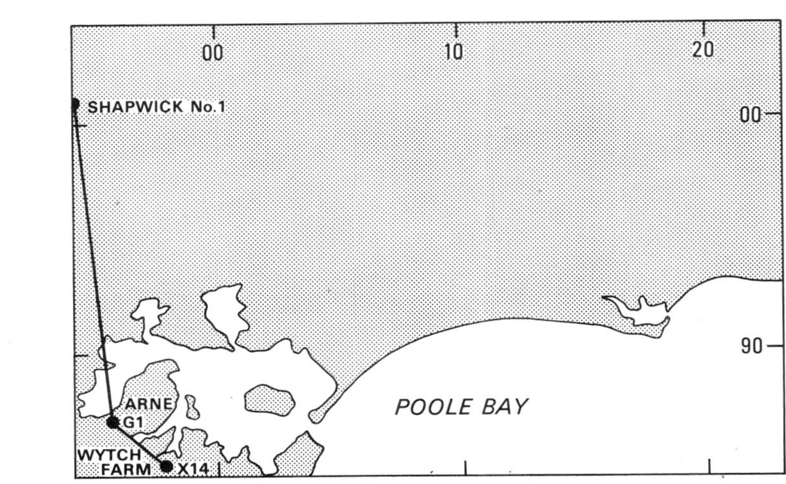

Figure 6 Gamma-ray and sonic velocity for oil wells X14, Arne G1 and Shapwick No. 1.

200 m. It also thins, but less rapidly, to the south across the South Dorset High. Within the Wytch Farm area, the Lower, Middle and Upper Lias maintain relatively constant thicknesses of 97 to 107 m, 50 to 63 m and 132 to 138 m respectively, and it appears that, with the exception of the Upper Lias, their constituent subdivisions behave similarly. There is some indication that the subdivisions of the Upper Lias, the Downcliff Clay and the Bridport Sands, show complementary thickness variations (Figure 7), though this relationship has not been studied in detail and the structural setting is not understood. The Lias is thought to have been deposited in open-marine shelf conditions, ranging from outer to inner shelf regions, which were subject to periodic shoaling, giving rise to the development of marine barrier sands such as the Bridport Sands (Knox et al., 1982).

The characteristic geophysical log signature (Figure 6) enables the Lias to be correlated regionally (Whittaker, 1985), and demonstrates that in the district it falls into the standard stratigraphical framework established in the type coastal areas (Cope et al., 1980). The base of the Jurassic System is taken at the lowest occurrence of the ammonite *Psiloceras planorbis* and probably lies some 3 to 4 m above the base of the Lias in the Wytch Farm area and elsewhere in the Bournemouth district.

The main lithostratigraphical units are as follows:

Lower Lias

The Lower Lias comprises a series of alternating medium to dark grey, locally silty and calcareous mudstones with interbeds of thin, pale to medium grey, hard, microcrystalline, variably argillaceous limestone.

These alternations in lithology give rise to the five lithostratigraphical subdivisions (Figure 6) which are widespread at both outcrop and subcrop in Dorset. They are readily identified by the large-scale cyclic pattern of their serrated log profiles (Whittaker et al., 1985). At Wytch Farm, they are represented by typical lithologies, apart from the Belemnite Marls, which are more calcareous than at outcrop.

Middle Lias

The Middle Lias is composed of pale grey, slightly silty, finely micaceous, locally carbonaceous and calcareous mudstones, interbedded with thin siltstones which become more prominent towards the top, where sporadic limestone ribs are taken to be the equivalent of the Marlstone Rock Bed. The geophysical log profile is correspondingly finely serrated and shows a gradual upward decrease in gamma-ray values to the top of the Middle Lias, where a marked decrease is taken to represent the rapid passage to the overlying beds. In some boreholes (e.g. Wytch Farm No. X14), a prominent sonic spike indicates the uppermost bed of the Middle Lias (Figure 6), but only in Wytch Farm No.F23 [SZ 0104 8514)] is there a prominent sand body at this horizon.

Upper Lias

The Upper Lias consists of the Downcliff Clay below and the Bridport Sands above. The Downcliff Clay is composed of medium grey, soft, calcareous, locally silty and micaceous mudstones, which become increasingly silty and micaceous upwards. Towards the top, pale grey, firm to friable, very fine-grained, moderately sorted, calcareous sandstones form a passage up into the Bridport Sands. In some boreholes, however, the change to the overlying formation is much more abrupt (e.g. Wytch Farm No.X14) (Figure 6). The geophysical log signature of the Downcliff Clay is finely serrated with high gamma-ray values towards the base.

The Bridport Sands, the less productive of the two oil reservoirs at Wytch Farm, are composed of white to pale grey, very fine-grained, moderately sorted sandstones containing subangular to subrounded grains, which grade into coarse-grained siltstones. The sandstones are feldspathic, locally glauconitic, micaceous and argillaceous, with scattered shell fragments which are locally concentrated to form bioclastic limestone beds. Their most striking feature is the alternation of hard, well-cemented, calcareous beds of low porosity with more friable, poorly cemented layers of greater porosity. The finer-grained interbeds are characteristically bioturbated. There is an upward increase in grain size, and an increase in the contrast between harder and softer beds, marking a fairly rapid upward passage from the Downcliff Clay. The geophysical log signature is characteristically highly serrated, reflecting the alternation of hard and soft beds, and has low gamma-ray and high sonic velocity values (Knox et al., 1982; Whittaker et al., 1985). Nevertheless, gamma-ray values are only moderately low, owing, it is presumed, to the feldspathic nature of the formation. Higher gamma-ray values tend to characterise the uppermost and lowermost parts of the signature where the formation tends to be more argillaceous.

MIDDLE JURASSIC

Inferior Oolite Group

The hard limestones of the Inferior Oolite Group thicken northwards into the Winterborne Kingston Trough, from a thin 3 m-thick sequence at Wytch Farm, to a maximum of 40 m against the Cranborne Fault (Whittaker, 1985, map 12). The Wytch Farm area has the thinnest Inferior Oolite in southern England and, because it occurs entirely in the typical Inferior Oolite 'condensed bed' facies, is thought to have been deposited on an offshore submarine swell (Gatrall et al., 1972; Penn et al., 1980; Penn, 1982). It constitutes some or all of the Bajocian [including Aalenian] Stage.

At Wytch Farm, the Inferior Oolite Group comprises grey to brown, coarsely shelly, 'pseudoconglomeratic' limestone, with large fragments of bivalves as well as corals and belemnites. Phosphatised hardgrounds are present, and contain large cobbles of algal limestone (up to 8 cm across), algal-coated limestone pebbles and limonitic ooliths, which are commonly set in a matrix of dark grey, pyritised, *Chondrites*-mottled silt. The geophysical log signature is distinctive in the Wytch Farm area, because the phosphatised limestone gives rise to pronounced low value sonic and high value gamma-ray spikes (Figure 6). It is thought that, by comparison with the cored sequence in the Winterborne Kingston Borehole (Penn, 1982, fig. 1), the high values are caused by minerals associated with hardground formation.

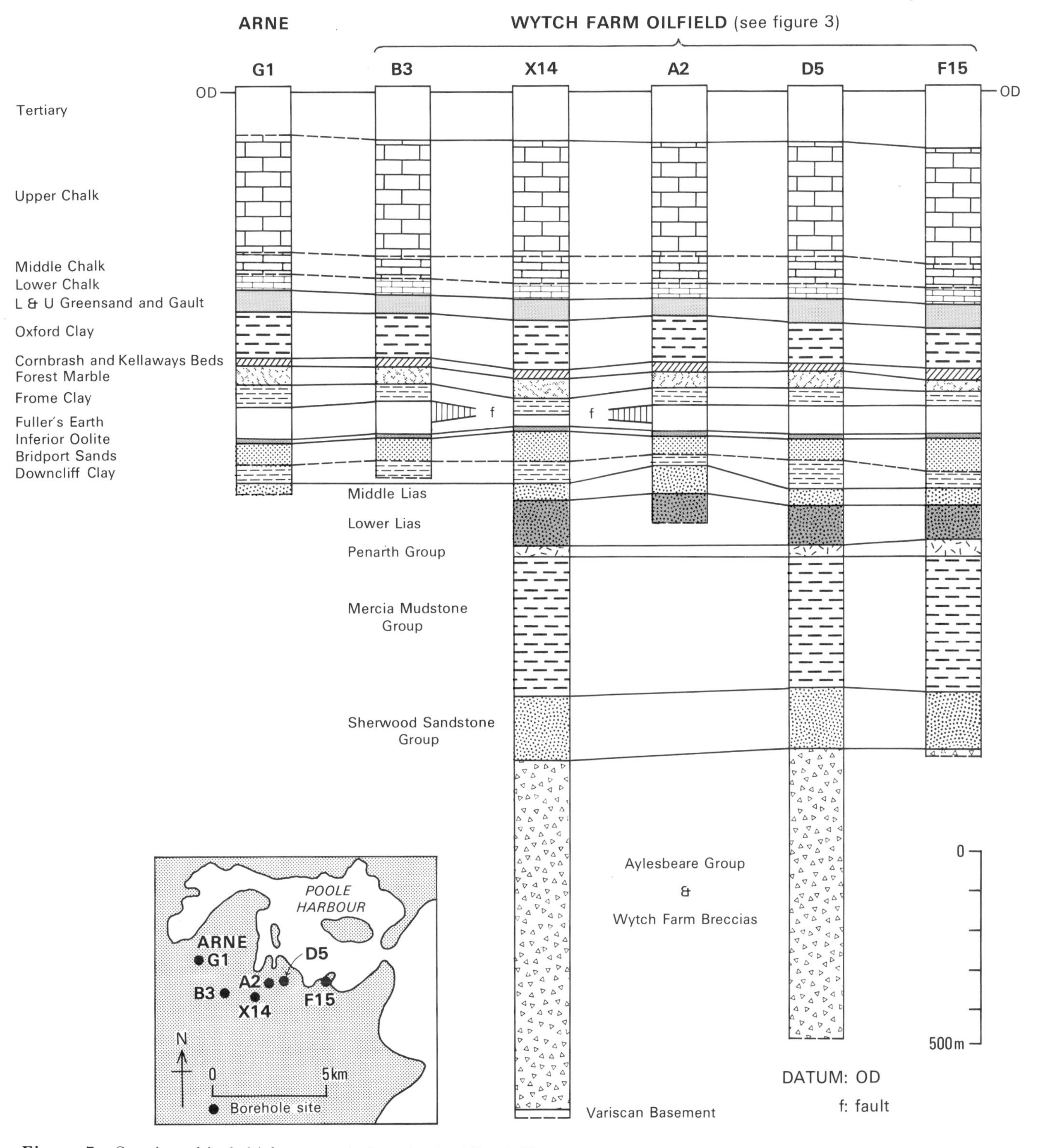

Figure 7 Stratigraphical thickness variations in the Wytch Farm area.

Great Oolite Group

The Great Oolite Group comprises, in ascending sequence, the Fuller's Earth, Frome Clay, Forest Marble and Cornbrash formations. It is a dominantly calcareous mudstone sequence with minor, thin, sporadically shelly limestones, siltstones and sandstones. At Wytch Farm, cores show a range of lithologies typical of the mudstone facies of the Great Oolite Group (Penn, 1982). Regional considerations, however, suggest that the argillaceous sequence passes north-eastwards into one dominated by limestones, and that this transition takes place, in part at least, within the Bournemouth district (Penn, 1985). The dominantly argillaceous nature of the group gives rise to a subdued, serrated gamma-ray log profile, the average gamma-ray value of which is lower than that of the underlying Lias (including the latter's sandy and feldspathic upper part) and the overlying Upper Jurassic clays. The sonic curve is moderately and characteristically serrated, in places deeply so (Figure 6), enabling the principal formations to be traced throughout the Wytch Farm area (Figure 6), and to be correlated with the fully cored succession in the Winterborne Kingston Borehole (Penn, 1982).

The Great Oolite Group is about 170 m thick around Wytch Farm but thickens north-westward along the axis of the Winterborne Kingston Trough to over 220 m; it also thickens north-eastwards. However, beyond the Cranborne Fault, on the Cranborne–Fordingbridge High, it thins to less than 120 m (Whittaker, 1985, map 14). The sediments were laid down largely as marine inner shelf muds in an area subject to cyclic shallowing. They form a major upward-shallowing sequence, such that the uppermost parts of the Forest Marble were deposited in subtidal to lagoonal conditions, subject to localised emergence and supply of terrigenous sediment (Penn, 1982).

As in the Winterborne Kingston Borehole, the Great Oolite Group of the district corresponds approximately to the Bathonian Stage, and the Upper/Lower Cornbrash boundary marks the Bathonian/Callovian stage boundary, the top of the Middle Jurassic Series in BGS practice.

Fuller's Earth

The Fuller's Earth comprises an interbedded series of pale to medium grey, variably calcareous, blocky to subfissile mudstones with sporadic carbonaceous material, interbedded with carbonate siltstones and pale grey, hard, argillaceous limestones. A prominent median limestone (1 to 2 m thick), the Fuller's Earth Rock, some 27 to 38 m above the top of the Inferior Oolite, separates the Lower and Upper Fuller's Earth. The majority of the subdivisions of the Lower Fuller's Earth (Units 1 to 9 of Penn et al., 1979) can be readily identified by comparison with the sequence in the Winterborne Kingston Borehole (Penn, 1982). The basal units (1 and 2) are thin and extremely condensed in most of the boreholes (e.g. Wytch Farm Nos B3, A6, A13, B21).

The Upper Fuller's Earth is between 2 m and 34 m thick and locally shows indications of attenuation in its lowest few metres. It gives rise to a sonic log profile which is characteristically deeply serrated by comparison with most of the Great Oolite Group (Whittaker et al., 1985). The low velocity values are taken to correspond to dark grey to black, ?water-retaining mudstones and shales found at this level in the type area (Penn et al., 1979).

Frome Clay

The Frome Clay comprises a lower, interbedded shelly limestone and shale unit (correlative of the Wattonensis Beds and contiguous strata in Winterborne Kingston Borehole) and an upper, uniform unit of variably calcareous mudstones. The median to lower portion of the lower unit was cored in the Wytch Farm No. F19 Borehole [SZ 0104 8574] where it is a dark grey grit composed entirely of comminuted shell debris set in a dark grey shaly matrix. Sporadic wisps and nodules of pale grey to white, microcrystalline shell-fragmental limestone form harder ribs, which are similar to beds associated with the two *Liostrea hebridica* lumachelles of the Dorset coast, the lower of which rests there on the Wattonensis Beds (Whittaker et al., 1985). The geophysical log profile of the lower part of this unit in Borehole F19 is characterised by low gamma-ray values and high sonic velocity. Although subdued, the finely serrated nature of the sonic log picks out the lamination of the grits.

When traced westwards from Wytch Farm No. F19 Borehole, the lower part of the Wattonensis Beds develops well-marked shaly interbeds which give rise to a more deeply serrated log profile. In particular, a mudstone interbed separates the lower part of the limestone sequence from the upper in some boreholes (e.g. D5 [SY 9947 8565] and Arne No. G1 [SY 9575 8704]).

The upper part of the lower unit is less argillaceous and more massive, as evidenced by lower gamma-ray values, which form a well-marked, broad gamma-ray low towards the top of the limestone sequence. Westwards from Borehole F19, the upper part passes into a pale grey to white, hard, micritic to coarse-grained siltstone and carbonate sand limestone, with common shell fragments, giving rise to a well-marked sonic spike and gamma-ray low on the geophysical log profile.

The limestones are overlain by distinctive dark grey, shaly mudstones. Above these 'black beds', the sequence becomes gradually paler in colour and increasingly calcareous and silty to the top of the Frome Clay. This upward change is recorded in the gradually upward decreasing gamma-ray values, increasing sonic velocity and somewhat stronger serrations of the log curves that make the geophysical log profile (Figure 6) of the Frome Clay one of the most distinctive in the Wessex Basin. In the more north-easterly of the Wytch Farm boreholes (e.g. Borehole F19), the slope of this distinctive log profile steepens (particularly that of the gamma-ray profile), suggesting that the beds become more calcareous towards the north-east. In the most easterly borehole (No.F23), three units of a white to yellowish brown, hard, sporadically oolitic limestone, in ascending sequence 5 m, 3 m, and 1.5 m thick, occur within the 24 m or so of these beds. These limestones give rise to very subdued serrated log curves of very low gamma-ray values and high sonic velocity. Thus it appears that the oldest of the Great Oolite Group argillaceous formations to pass laterally north-eastwards into the limestone facies, when traced from basin to shelf, is the Frome Clay (cf. Penn et al., 1979). It is probable that a substantial part of the Bournemouth district is underlain by these limestones, which correspond to the Great Oolite Formation of the Bath area.

The Frome Clay thickens from c.46 m in the east of the district to 58 m in Arne No.1 Borehole, as it becomes increasingly argillaceous.

Forest Marble

The Forest Marble is a sequence of pale grey to green, calcareous, poorly bedded mudstones, interbedded with thin bioclastic limestones and shell-beds. The mudstones include wispy bedded siltstones and irregularly bedded, very fine- to fine-grained, calcareous sandstones which commonly contain scattered carbonaceous debris. The thickness is 47 to 48 m throughout the Wytch Farm area.

The geophysical log signature is typically spiky, with high sonic and low gamma-ray values characterising the limestones and sandstones (Figure 6). It is sufficiently distinctive to be matched in detail with that obtained from the fully cored Winterborne Kingston Borehole, thus enabling shell beds such as the Boueti Bed and the Digona Bed to be identified (see Penn, 1982; Whittaker et al., 1985).

Cornbrash

The Cornbrash can be divided into two members, the Lower Cornbrash and the Upper Cornbrash; the latter, strictly, forms the basal unit of the Upper Jurassic sequence. However, it is described in this section for convenience.

The formation comprises a sequence of pale grey, shell-fragmental limestones, interbedded with dark grey, shell-grit mudstones. Samples of core material show micritic, argillaceous, wispy bedded limestones with bioturbated shell-debris, typical of the Lower Cornbrash, overlain by coarse-grained, sandy limestone beds of the Upper Cornbrash, which become more argillaceous upwards.

The formation maintains a constant 10 m thickness throughout the Wytch Farm area, approximately equally divided between Upper and Lower members in the east. Towards the west, however, the Lower Cornbrash comprises about one third of the formation.

The geophysical log traces (Figure 6) are characterised by low gamma-ray and high sonic velocity values. Generally, the Upper Cornbrash may be distinguished from the Lower Cornbrash because the latter has a less deeply serrated gamma-ray profile of lower values, the bell-shaped profile of which reflects an upward passage to Kellaways Clay.

UPPER JURASSIC

The Upper Jurassic succession is between 100 and 200 m thick and was extensively affected by erosion prior to the deposition of the Cretaceous. Only the lowest formation, the Kellaways Beds, is preserved completely throughout the district, and hence the thickness variation of individual Upper Jurassic formations is difficult to evaluate. The Lower Cretaceous sequence rests on the Oxford Clay in the oilfield area, approximately delineating the South Dorset High (Figure 21). Over much of the rest of the district, the Lower Cretaceous sediments overlie Corallian Beds and Kimmeridge Clay.

The Upper Jurassic clay sequence was deposited in a shelf sea environment which was subject to cyclic shallowing. Where shallowing was accompanied by the supply of coarse clastic sediment, barrier sands such as the Kellaways Sand were formed; where, however, there was no influx of terrigenous sediment, typical carbonate-platform sediments such as the Corallian Beds were generated. Periods of restricted circulation gave rise to deoxygenated bottom waters and to the deposition of the organic-rich shales of the Oxford Clay and Kimmeridge Clay.

Kellaways Beds

The lower part of the Kellaways Beds, the Kellaways Clay, comprises pale to dark grey, slightly silty and micaceous, calcareous, commonly bioturbated shelly mudstones. These beds gradually pass upwards through medium grey siltstones interbedded with mudstone and calcareous sandy lenses, to dark grey, coarse-grained silt, and fine-grained, calcareous and argillaceous, sporadically shelly sandstones. The sandy upper part, the Kellaways Sand, locally contains white to dark grey, hard, silty limestone doggers. The Kellaways Beds maintain a uniform thickness of 13 m throughout the Wytch Farm area.

The geophysical log signature is typically funnel-shaped (Figure 6), with an upward increase in gamma-ray values accompanied by increased sonic velocities, reflecting the upward increase in sand content. Prominent sonic spikes and minor gamma-ray lows in the Kellaways Sand interval correspond to well-cemented limestones. The formation is commonly capped by such a limestone, which gives rise to a prominent flat top to the log signature, below that of the Oxford Clay.

Oxford Clay

The Oxford Clay in the Wytch Farm area is a predominantly argillaceous sequence in which grey to brown, subfissile, calcareous, micaceous, shelly and slightly carbonaceous mudstones, interbedded with thin siltstones (the Lower Oxford Clay), pass upwards into paler grey, soft, calcareous, pyritic mudstones interbedded with thin, fissile, slightly silty and calcareous siltstones and white to pale brown argillaceous limestones of the Middle and Upper Oxford Clay. The geophysical log signature (Figure 6) is finely serrated and of high gamma-ray and low sonic velocity values in the argillaceous parts of the sequence, but is characterised by discrete spikes where silty and calcareous beds predominate. These silty and calcareous beds tend to occur as part of upward-coarsening cycles and mark the top of the Lower Oxford Clay about 30 to 40 m above the base, and the top of the Middle Oxford Clay about 80 m above the base. The formation ranges in thickness from about 100 to 118 depending on the extent of pre-Cretaceous erosion.

Corallian Beds

The Corallian Beds occur at subcrop beneath Cretaceous strata on the Cranborne–Fordingbridge High in the north-east of the district, and on the South Dorset High in the Wytch Farm area. Some 30 m of Corallian Beds lie in faulted contact with the Oxford Clay in the Wytch Farm No. 2 Borehole [SY 9895 8555]. In the intervening ground, they occur beneath thin Kimmeridge Clay in the east-south-east extension of the Winterborne Kingston Trough.

The Corallian Beds comprise a lower succession of pale to dark grey, firm to hard, very calcareous, locally carbonaceous, slightly pyritic, finely micaceous and subfissile mudstones with abundant shell fragments, probably equivalent to the Nothe Clay of the Dorset coast (Whittaker et al., 1985). These pass upwards into a grey, medium- to coarse-grained, moderately sorted sandstone containing subangular to rounded grains, which is sporadically argillaceous and locally cemented by calcite, probably the correlative of the Bencliff Grit in the Winterborne Kingston Borehole. This in turn passes up into an interbedded sequence of white to cream, oolitic and pisolitic limestones, and pale to medium grey calcareous and very silty mudstones, probably equivalent to the Osmington Oolite in Winterborne Kingston Borehole.

The argillaceous lower part of the Corallian Beds gives rise to high gamma-ray and low sonic velocity values. The finely serrated geophysical log traces are sporadically punctuated by sonic spikes and gamma-ray lows corresponding to discrete, thin argillaceous limestones. The overlying sandstones give rise to lower gamma-ray values and interval transit times, and a more serrated log signature. The coarsening-upward signature is completed by the low gamma-ray and high sonic velocity values associated with the limestones present near the top of the Corallian Beds. The interbedded mudstones are sufficiently calcareous and silty to give rise to moderate to poorly serrated log profiles. These beds have a log signature resembling that of the Corallian Beds of Winterborne Kingston Borehole (Whittaker et al., 1985), supporting the suggested correlations.

'Ampthill Clay Equivalent'

Overlying the Corallian Beds in Wytch Farm No. A2 Borehole [SY 9895 8555] are 6.4 m of pale to medium grey, calcareous, very silty mudstones with sporadic thin argillaceous limestones. The gamma-ray profile of these beds is moderately serrated and shows an upward increase in values. In Shapwick No.1 Borehole, the core between 492.9 and 494.7 m proved pale to medium grey, very silty, calcareous mudstones with chondritic mottling, pyritised trails and pins, and pyrite staining. From these a bivalve fauna, including *Grammatodon, Thracia,* nuculoids, oysters and pectinids was obtained (determinations by Dr B M Cox). These beds give rise to moderately serrated geophysical log curves in which gamma-ray values are markedly less than those of the Kimmeridge Clay (Figure 6). The lithology of the mudstones and their geophysical log signature resemble those of the mudstones which lie between the Corallian Beds and the Kimmeridge Clay in the Winterborne Kingston Borehole, where they have been termed 'Ampthill Clay Equivalent' (Rhys et al., 1982), and are considered to lie within the uppermost Oxfordian Stage.

Kimmeridge Clay

Kimmeridge Clay is preserved beneath the Cretaceous overstep in the central region of the district. On the Cranborne–Fordingbridge High to the north-east, and in the Wytch Farm area, however, it has been removed by Cretaceous erosion. Some 60 m were drilled in Shapwick No.1 Borehole, where core from the top of the formation (428.9 m to 431.9 m) is of pale to medium grey mudstone interbedded with dark grey oil-shale, overlain by similar, but silty and sporadically calcite-cemented, mudstone. The fauna from this interval (determinations by Dr B M Cox) includes the brachiopod *Lingula*; the gastropod *Dicroloma*; bivalves including *Entolium*, *Grammatodon*, *'Lucina'*, *Nanogyra virgula*, common nuculoids, *Oxytoma* and *Protocardia*, and a fauna of ammonites which include *Amoeboceras (Nannocardioceras)*, *Aspidoceras* (and *Laevaptychus*), *Aulacostephanus* and *Sutneria*. This lithology and fauna indicate beds 29 and 30 of the standard Kimmeridge Clay sequence established by Cox and Gallois (1981). Similar fossiliferous mudstones and shales below are interbedded with thin, medium grey, calcite-veined limestones. An abundance of crinoid ossicles at 454.2 m to 457.2 m depth probably indicates Bed 18 of the standard sequence.

Beds 29–30 of the Kimmeridge Clay occur in the middle of the Eudoxus Zone (Cox, 1982) of the Kimmeridgian Stage, and Bed 18 indicates a horizon high in the Mutabilis Zone. By comparison with the succession in the Winterborne Kingston Borehole (Whittaker et al., 1985), where the Lower Cretaceous beds rest on a horizon low in the Eudoxus Zone, it appears that the Eudoxus, Mutabilis, Cymodoce and Baylei Zones are respectively 20.9+, 32, 5 and 1.7 m thick. The geophysical log signature is typically rhythmic in nature and finely serrated (Figure 6).

CRETACEOUS

The Lower Cretaceous formations, totalling 100 m in thickness, were laid down in an open shelf sea which was subjected to cyclic shallowing leading to the deposition of the marginal sandy facies of the Lower and Upper Greensand.

Lower Greensand

The Lower Greensand is thought to rest in pockets on the eroded Jurassic surface, because it has been proved only in Arne No. G1 Borehole. There, some 3 m of pale grey and green, fine-grained, calcareous sandstone rest on Upper Oxford Clay. The Lower Greensand gives rise to a single sonic velocity peak and a corresponding gamma-ray low.

Gault

Some 20 to 35 m of Gault have been proved in all the Wytch Farm boreholes and in Shapwick No. 1 Borehole. It occurs throughout the Bournemouth district and consists of medium to dark grey, poorly fissile mudstones and shales, with sporadic very fine quartz and glauconite grains, and a fauna of serpulids and crushed, pyritised bivalves. At the base there is a yellowish green, medium-grained, sandy, glauconitic clay, which becomes increasingly silty and sandy upwards. By comparison with the cored sequence in Winterborne Kingston Borehole (Morter, 1982), the sandy basal facies of the Gault corresponds to the Basement Bed, and the argillaceous part to the Foxmould in that borehole. The Basement Bed of the Gault probably belongs to the early Albian Substage, and the remainder of the Gault to the mid Albian Substage (Morter, 1982).

The geophysical log signature (Figure 6) is characterised

by a poorly serrated curve of high gamma-ray values and low sonic velocity. This is punctuated by high velocity sonic spikes and more subdued gamma-ray lows, the response of the silty and sandier beds. The main feature of the signature is the overall upward decreasing gamma-ray values, which reflect the upward coarsening of grain size as the beds become more silty and sandy. No such change takes place in sonic velocity, which is taken to indicate that the sandy and siltier levels are weakly consolidated; probably only the well-cemented layers give rise to well-marked sonic spikes.

Upper Greensand

The Upper Greensand occurs throughout the district. It comprises some 29 m to 39 m of white to pale grey, firm to friable, very fine- to medium-grained, slightly argillaceous and pyritic, well-sorted sandstone with green glauconite specks; it becomes siltier towards the base and has a prominent dark greenish grey fine-grained, sandy and glauconitic mudstone in its median part.

The gamma-ray log profile (Figure 6) is characterised by a moderate to deeply serrated curve, reflecting the variably argillaceous fine- to medium-grained sandstones. The profile shows a decrease in values upwards as the succession becomes sandier and coarser grained. The sonic curve is characterised by low values punctuated by high velocity sonic spikes, showing that the sandstones are poorly consolidated, with sporadic well-cemented levels. Superimposed on the upward decrease in gamma-ray values is a persistent smaller scale rise in values about the middle of the formation, which gives a characteristic 'pinched' gamma-ray profile. The Upper Greensand belongs in the late Albian Substage (Morter, 1982, fig.1).

Chalk

About 415 m of Chalk have been proved in the Wytch Farm area, and it is likely that a similar thickness underlies the entire district (Whittaker, 1985, map 25). Lower, Middle and Upper Chalk are readily distinguished on geophysical logs but, because the Upper Chalk crops out in the north-west near Sturminster Marshall, only Lower and Middle Chalk are included in this chapter.

Lower Chalk

The Lower Chalk consists of off-white, pale grey, bluish grey to medium grey, firm to soft, silty textured, marly chalk. It is markedly silty and sandy (with sporadic glauconite) near the base. A prominent condensed basal bed marks the base of the middle Cenomanian, which rests disconformably on the late Albian Upper Greensand. This bed consists of either hard, sandy and glauconitic clay, the Glauconitic Marl, or marly chalk with patches of pinkish clay and rare 'flint' pebbles, associated with a greenish black, glauconitic, fine- to coarse-grained sand. The top of the hard bed in the middle of the Lower Chalk corresponds to the middle/upper Cenomanian boundary. The top of the Lower Chalk comprises interbedded marls and marly chalks, the Plenus Marls. The Lower Chalk is 44 to 48 m thick in the Wytch Farm area (Figure 7), thickening uniformly to 56 m in Shapwick No.1 Borehole (Figure 6).

The geophysical log signature (Figure 6) is characteristic and compares closely with the cored succession in Winterborne Kingston Borehole (Wood et al., 1982; Lott, 1982). It consists of a very subdued, moderately serrated gamma-ray profile, composed of very low values which decrease further in value upwards as the clay content decreases. The sonic curve is more strongly, but only moderately serrated (representing alternations between harder chalks and softer more argillaceous chalks), and shows a complementary upward increase in sonic velocity. Superimposed on this overall pattern are widespread, though stratigraphically restricted variations which enable marker beds to be recognised and allow further subdivision of the sequence. Thus the basal bed is marked by a gamma-ray peak and high velocity sonic 'spike', and the Plenus Marls at the top of the sequence are marked by a prominent, though moderate, gamma-ray peak and a more prominent low velocity level, which rapidly increases towards the top of the Plenus Marls. In addition, a broader sonic peak usually located about a half to a third of the way up the Lower Chalk, marks a persistent level of hard chalk which separates the more argillaceous Chalk Marl beneath from the Grey Chalk above.

Middle Chalk

The Middle Chalk consists of pale grey, greyish white to white, firm to moderately hard, silty textured chalk with sporadic thin marl seams and hardground beds. Pale grey to pale brown flints are present in its upper part.

The thickness of the Middle Chalk is consistent throughout the district (Figure 7); depending on where its upper boundary is taken, as outlined below, it ranges from 40 to 50 m, or 50 to 60 m in thickness.

The geophysical log signature (Figure 6) is characterised by a poorly serrated gamma-ray curve registering uniformly very low values, and a moderately serrated sonic log profile of moderate to low velocity, reflecting the alternation of harder and softer chalks. Minor gamma-ray peaks and more prominent low velocity sonic values correspond to marl bands, which can be traced throughout the district. Persistent high sonic velocity peaks are the response of hard chalks, commonly the result of hardground formation, which can be similarly traced. Of these, the basal Melbourn Rock is readily identifiable by its proximity to the Plenus Marls, but the higher Spurious Chalk Rock and underlying Glynde Marl are more readily distinguishable by their dramatic contrast in sonic velocity. Difficulty, however, is encountered in locating the Chalk Rock (and hence the Middle/Upper Chalk boundary) because of its somewhat undistinguished log character and the absence of core material, both within the district and nearby. Wood et al. (1982) assert that it lies at a level which at Wytch Farm would be some 10 m above the Glynde Marl, but oil-company practice places it some 10 m higher.

THREE

Cretaceous: Upper Chalk

The Upper Chalk is a firm white fossiliferous chalk, with scattered layers of flint nodules and thin marl seams. Eight macrofaunal assemblage biozones are identified in southern England, of which the two highest, the *Gonioteuthis quadrata* Zone, succeeded by the *Belemnitella mucronata* Zone, are represented at outcrop within the Bournemouth district. Mortimore (1986), working largely on the Chalk of Kent and Sussex, erected a new lithostratigraphic unit, the Sussex White Chalk Formation, to embrace both the Middle and Upper Chalk of the traditional classification. It comprises six members, each subdivided into a number of beds. Of the newly recognised members, the two highest, the Culver Chalk Member below and the Portsdown Chalk Member above, approximate to strata of *quadrata* and *mucronata* zone age respectively[1]. Gale et al. (1987) proposed a further new name, the Studland Chalk Member, for the flinty, marl-free chalks that characterise the higher part of Mortimore's Portsdown Chalk Member.

In the adjacent Dorchester district (Sheet 328) to the west, strata of both *quadrata* and *mucronata* zone age correspond to dip and scarp features within the Upper Chalk (Bristow, 1987a), and were thought by Bristow (1987a) to equate with the Culver and Portsdown Chalk members of Mortimore (1986). Unfortunately, in the absence of suitable sections thoughout the Upper Chalk in the Bournemouth district, the exact stratigraphical details of the presumed Culver and Portsdown Chalk members were not established. However, recent mapping west and north-west of the Bournemouth district by Dr C M Barton, and a re-examination of the BGS palaeontological material from that area by Mr C J Wood shows that the pronounced, stratigraphically lower scarp in strata of *quadrata* Zone age in the Dorchester and Ringwood areas (Bristow, 1987a), which is not present at outcrop in the Bournemouth district, probably corresponds to the relatively coarse-grained chalks comprising the Sompting Beds within the Culver Chalk Member, and that the strata fall within the *Applinocrinus cretaceus* Subzone of the *quadrata* Zone (Wood *in* Barton, 1990). The dip slope developed on strata above the feature occurs in the north-west of the district, but there is no exposure. The scarp associated with strata of *mucronata* Zone age crosses the district. The feature is only weakly developed where it emerges from beneath Quaternary deposits of the River Stour, near Shapwick [ST 948 020], but it becomes more pronounced towards the east near Badbury Rings [ST 960 034], and is about 10 m high where it leaves the district north-east of King Down Farm [ST 975 042]. The lower part of the scarp is probably formed by the more marly chalks of the Portsdown Chalk Member (as restricted by Gale et al., 1987), and is capped by the harder, more flinty beds of the Studland Chalk Member. On the adjacent Dorchester (328) Sheet, a section in the middle of the scarp face in the presumed Portsdown Chalk Member reveals a 5 cm-thick marl (probably the Shide Marl seam) (Bailey et al., 1983; Mortimore, 1986) in the middle of strata of *mucronata* Zone age (Bristow, 1987a). However, a pit [9493 0192] in *mucronata* Zone strata low down the scarp in the present district revealed no marl seam. All other sections in the district presumably fall within the Studland Chalk Member. All the pits fall within the Pre-Weybourne Chalk of the Norfolk sequence on the basis of the occurrence of *Echinocorys* ex gr. *conica*.

Whilst it is possible, at least within a limited area of the Bournemouth and adjacent districts, to map lithostratigraphical divisions that have previously been identified only in quarry and cliff sections in areas to the east, these divisions are not shown on the published 1:50 000 map because only a small area of Upper Chalk has been mapped to date.

The base of the Upper Chalk in the Shapwick No.1 Borehole [ST 9429 0135] is drawn at a depth of 243.23 m by the oil company geologists, but there is no indication from the brief lithological log, nor from the gamma-ray and resistivity logs, why the boundary is placed at this level. The position of the borehole, which was uncored, on the dip slope of the Culver Chalk, indicates that it commenced in the upper part of the *quadrata* Zone. It proved 225 m of Upper Chalk beneath supposedly 18 m of drift. In addition, about 60 m of strata of late *quadrata* and *mucronata* Zone age crop out between the top of the Chalk in the borehole and the base of the Palaeogene strata. Strata of the *mucronata* Zone are calculated to be 35 m thick near Badbury Rings and 35 to 40 m thick north of Sturminster Marshall. They may be thicker farther south, where pre-Tertiary erosion was less severe. CRB

DETAILS

At Shapwick, a largely infilled pit [ST 9493 0192] exposes only 1.5 m of flaggy white chalk. Jukes-Browne (1904, p.122) recorded a 6 m section in this pit, in soft white chalk with a few scattered yellow-coated flints, together with '*Porosphaera globularis,* small *Ostrea* sp., *Belemnitella mucronata,* small ovate *Echinocorys* and *Salenia granulosa*'. White (1917) noted a slightly different section here, with three indistinct courses of small, thick-rinded flints and a few oblique veins of tabular flint, and recorded the following fossils: '*Onychocella lamarcki, Stomatopora granulata, Axogaster cretacea* and *Dimyodon nilssoni*'. From its position low down on the scarp face, this pit probably lies close to the base of the *mucronata* Zone.

At White Mill Bridge there is a large quarry [ST 959 008] in which Jukes-Browne (1904) described a 4.8 m section of soft white flintless chalk of *mucronata* Zone age; '*Belemnitella mucronata* and *Kingena lima*' were found there. The quarry has been considerably enlarged since Jukes-Browne's day, but is now (1987) being filled. Some 12 m of white, blocky, thinly bedded chalk in bedding units up to 10 cm thick can be seen; the dip is 2.5° SE. On the west side of the pit, about 5 m of chalk are exposed beneath a persistent layer of

1 At the type locality on the Isle of Wight, the Culver Chalk begins approximately 5 m above the base of the *quadrata* Zone and ends approximately 10 m below the top; the Portsdown Chalk thus includes the highest part of the *quadrata* Zone.

flints [ST 9588 0082]; on the east side, about 7 m of chalk occur above the flint bed [ST 9595 0062]. There are small (1 cm) to large (10 cm) rounded flints, elongate irregular nodular flints up to 30 by 10 cm, and cylindrical vertical flints up to 10 cm high by 2 cm diameter. Two *Belemnitella* were found below the flint bed, and *Porosphaera* sp., *Belemnitella* and a large *Echinocorys* (*E.* cf. *subglobosa*) above. Holmes (MS, BGS) found *Ophryaster oligoplax* in this pit.

Jukes-Browne (1904) calculated that 60 m of strata intervened between this pit and the base of the Tertiary strata at Badbury Rings. However, he made no allowance for dip, and was unaware of the presence of small Tertiary outliers less then 800 m from the quarry. With this new evidence, it is calculated that only 25 m of strata occur between the pit and the base of the Tertiary; the thickness of *mucronata* Zone strata below the pit is not known with certainty, but it may be 10 to 15 m.

A pit [ST 9694 0374] at King Down Farm, 10 m below the base of a Tertiary outlier, exposes 3 m of rubbly chalk. At the top of the section, there is an undulating tabular flint, 1 m long by 1 cm thick; one nodular flint was found in the middle of the section. Fossils include *Cretirhynchia woodwardi, Belemnitella mucronata* and *Echinocorys* ex gr. *conica* which indicate the lower part of the *mucronata* Zone (Pre-Weybourne Chalk).

White (1917) described another pit [ST 9758 0290] at about the same stratigraphical level, which was worked in two levels; the upper tier [ST 9735 0286] showed 3 m of soft flintless chalk with *Belemnitella mucronata*; the lower tier [ST 9759 0289] exposed 4.5 m of soft chalk with a few scattered small flints, and one well-marked undulating band of large flints with thick grey rinds near the top. Irregular shaped rusty pyrite nodules were plentiful. Fossils included: '*Porosphaera globularis, P. nuciformis, Serpula plexus, Rhynchonella plicatilis, Inoceramus inconstans, Metopaster parkinsoni, Mitraster rugatus, Pentagonaster quinquelobus, Cidaris hirudo, C. subvesiculosa?, Echinocorys scutatus* and *Helicodiadema fragile*' (White, 1917, p.7). Holmes (MS, BGS) also found *Kingena pentangulata, ?Arthraster cristatus* and *Echinocorys* possibly *E. lata fastigata.*

A third pit [ST 9785 0408] 1.2 km NNE, which lies close to the base of a Palaeogene outlier, showed a similar sequence to that in the lower tier described above (White, 1917).

A pit [ST 9812 0470] at Bradford Barrow shows 2.5 m of blocky weathering chalk, with a subhorizontal ferruginous-stained layer near the middle; small nodular flints occur. One specimen of *Echinocorys* ex gr. *conica* was found. Reid (1902) recorded *E. vulgaris* from this pit.

A pit [ST 998 032] at High Wood formerly exposed 3.6 m of chalk with scattered thick-skinned black flints and *Belemnitella mucronata* (Reid, 1902, p.7). This pit lies at the top of the Chalk sequence, since Palaeogene strata crop out immediately to the east and north. The Marl Pit [ST 9905 0215] 500 m ESE of Chilbridge Farm, has now been filled; up to 2.5 m of rubbly chalk was seen around the edges. Reid (MS map, BGS) noted scattered thick-rinded flints and iron pyrites, and Holmes (MS, BGS) found *Spondylus latus, B. mucronata* and *Metopaster tumidus* in this pit.

A pit [SU 0012 0486] in the extreme north-west of the district still exposes a 7 m section of soft, white to off-white chalk in 0.1 to 0.5 m-thick beds; the top 1 to 3 m are rubbly. Flints range from small (1 to 2 cm) and rounded, up to large (20 cm across) and horned, and are scattered irregularly throughout. Several yellow-stained joint faces cut the face. Two fragments of *Belemnitella mucronata* were found. Reid (1902, p.7) saw this pit when it was 10.7 m deep in massive soft chalk, and noted a few thick-skinned flints and *Belemnitella mucronata* near the base.

Belemnitella mucronata was found in the upper 30 m of chalk in a well [SU 0067 0093] at Wimborne (Whitaker and Edwards, 1926).

CRB

FOUR

Palaeogene: Reading Formation and London Clay

INTRODUCTION

The Palaeogene formations, which underlie most of the district, total around 420 m in thickness. Much of the higher part of the sequence is well exposed in the coastal sections between Poole Head and Barton on Sea. The Palaeogene strata rest unconformably on the eroded top of the Upper Chalk, but without marked angular discordance. The succession youngs generally eastwards.

The deposits consist almost exclusively of siliciclastic rocks ranging from clays to pebble beds, and were divided during the survey into formations and members on the basis of their dominant lithological characters (Figure 8). The sequence consists of a number of sedimentary cycles, each of which commenced with a marine transgression; some transgressive surfaces are marked by a thin bed of flint pebbles. The cycles have been used to define informal divisions in the London Clay (King, 1981) and the Bracklesham Group (Plint, 1983b). The stratigraphy shown on the present map differs considerably from that of the 1895 edition.

Selected stages in the evolution of the nomenclature of the Palaeogene rocks in the district are shown in Figure 9.

PALAEONTOLOGY

Attempts have been made to produce a biostratigraphical framework for the Palaeogene sediments of the Hampshire Basin, using dinoflagellate cysts, pollen and spores, foraminifera and molluscs. Unfortunately, during the deposition of the older Palaeogene sediments (London Clay and Bracklesham Group) in the western part of the Hampshire Basin, fully marine conditions rarely persisted and the sediments were commonly laid down in nearshore lagoonal, brackish or fluviatile environments. As a consequence, a biostratigraphical correlation with the more marine part of the Hampshire Basin to the east can only tentatively be made.

The West Park Farm Member at the base of the London Clay is unfossiliferous, except for oysters in its basal bed. The succeeding non-reddened beds of the London Clay are only sparsely fossiliferous; molluscs are found over only a limited area at the base of the Warmwell Farm Sand. The argillaceous parts of the London Clay have yielded a restricted assemblage of dinoflagellate cysts and foraminifera. The Bracklesham Group has yielded few macrofossils, but sparse dinoflagellate cysts have been recovered from the clay members. The Barton Group is commonly decalcified at the surface and few fossils are preserved, but where unweathered it is very fossiliferous at some levels, particularly in the Barton Clay and Chama Sand. The Headon Formation is shelly at some levels; the Lyndhurst Member contains a fauna of thick-shelled marine bivalves, and the remainder of the formation is dominated by fresh to brackish water molluscs.

The foraminifera of the Palaeogene sediments of the district are mainly benthonic, and their distribution is largely environmentally controlled; they are, therefore, of limited value for stratigraphical correlation. In the London Clay of the main part of the Hampshire Basin, Wright (1972) recognised an influx of several species of planktonic foraminifera which he called the 'Planktonic Datum'. Such a fauna has been found at only one locality in the district, near Knighton [SZ 0535 9789] (p.29). Species of the larger planktonic foraminifer *Nummulites* are important stratigraphically in the main part of the Hampshire Basin, but there is only one occurrence of the genus in the district, that of *N. prestwichianus* in the Barton Clay (p.5).

Many hundreds of molluscs have been recorded from the Barton Group, and attempts have been made (Burton, 1933) to use them to subdivide the group. However, most of the molluscan assemblages are facies controlled and are of limited biostratigraphical value.

Costa and Downie (1976) erected eight zones for the Tertiary of north-west Europe based on assemblages of species of the dinoflagellate cyst genus *Wetzeliella*. Independently, Eaton (1976) correlated the Bracklesham Group on the east and west coasts of the Isle of Wight, also using dinoflagellate cysts. Bujak et al. (1980) erected dinoflagellate cyst zones for the Barton Group. However, in the western part of the Hampshire Basin, because of the rapidly fluctuating marginal marine environments such as those in the London Clay and Bracklesham Group, these fossils are not so effective for detailed correlation. However, they have proved particularly important in establishing the presence of marine events within the dominantly fluviatile Poole Formation and Branksome Sand. Because of the more marine nature of the Barton Clay compared to the Bracklesham Group, the dinoflagellate cysts have proved more reliable for correlation.

READING FORMATION

At outcrop, strata formerly classified as 'Reading Beds' are now considered to be red-stained strata of the London Clay and have been renamed the West Park Farm Member (see below). Therefore, in the north-western part of the district, London Clay rests directly on the Chalk and the Reading Formation is missing. However, Reading Formation strata were proved at depth in the Christchurch [SZ 2002 9301] and Knapp Mill [SZ 1544 9380] boreholes, and in boreholes around Wytch Farm, in the southern part of the district. The Christchurch Borehole yielded cuttings of bright red clay with a varying silt content. The gamma-ray log responses from this part of the hole are similar to those characterising the Reading Formation in boreholes near Southampton, such as Bunkers Hill Borehole [SZ 3038 1498], and in Sandhills No.1 Borehole [SZ 4570 9085] on the Isle of Wight,

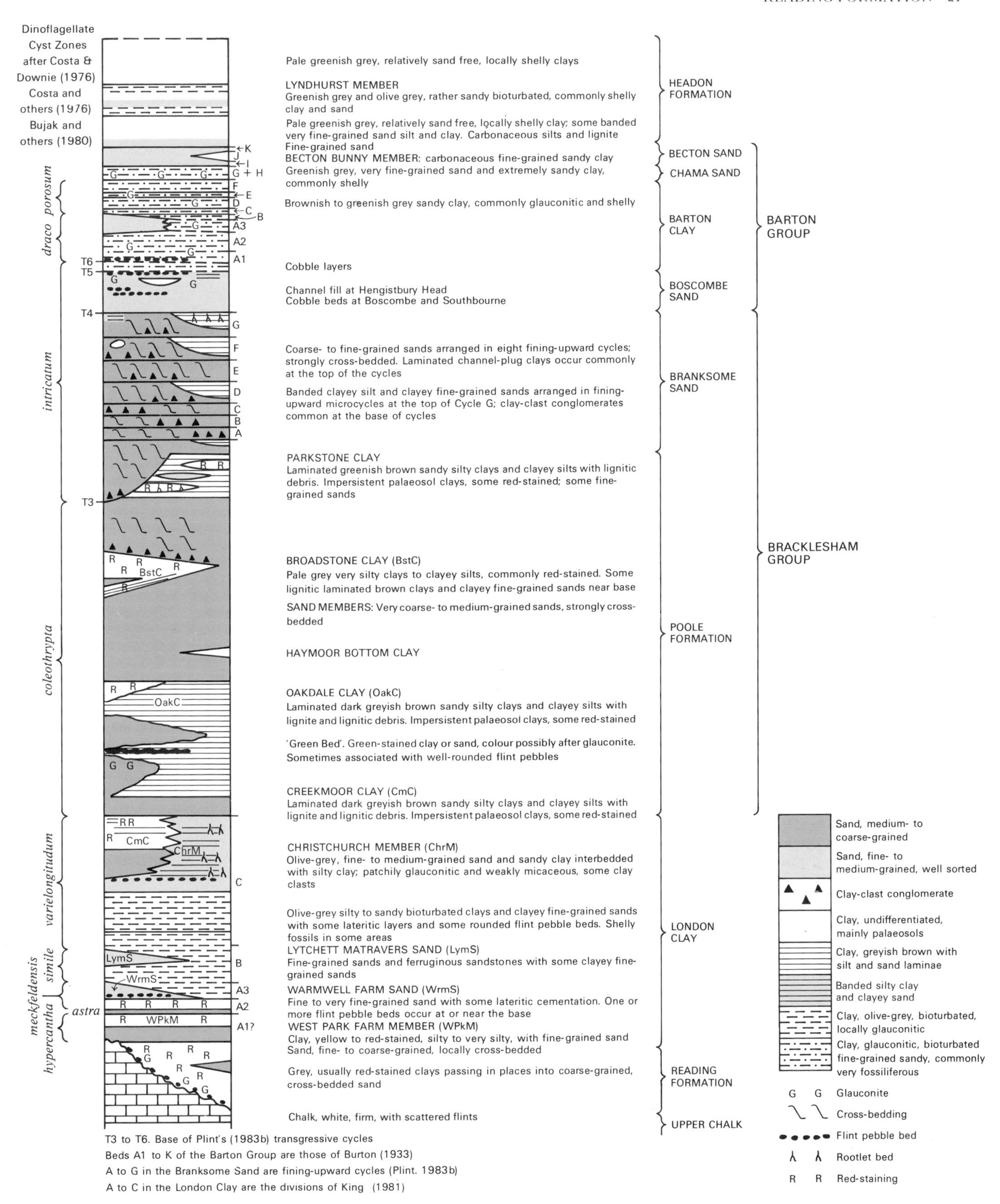

Figure 8 Generalised stratigraphy of the Palaeogene strata of the district.

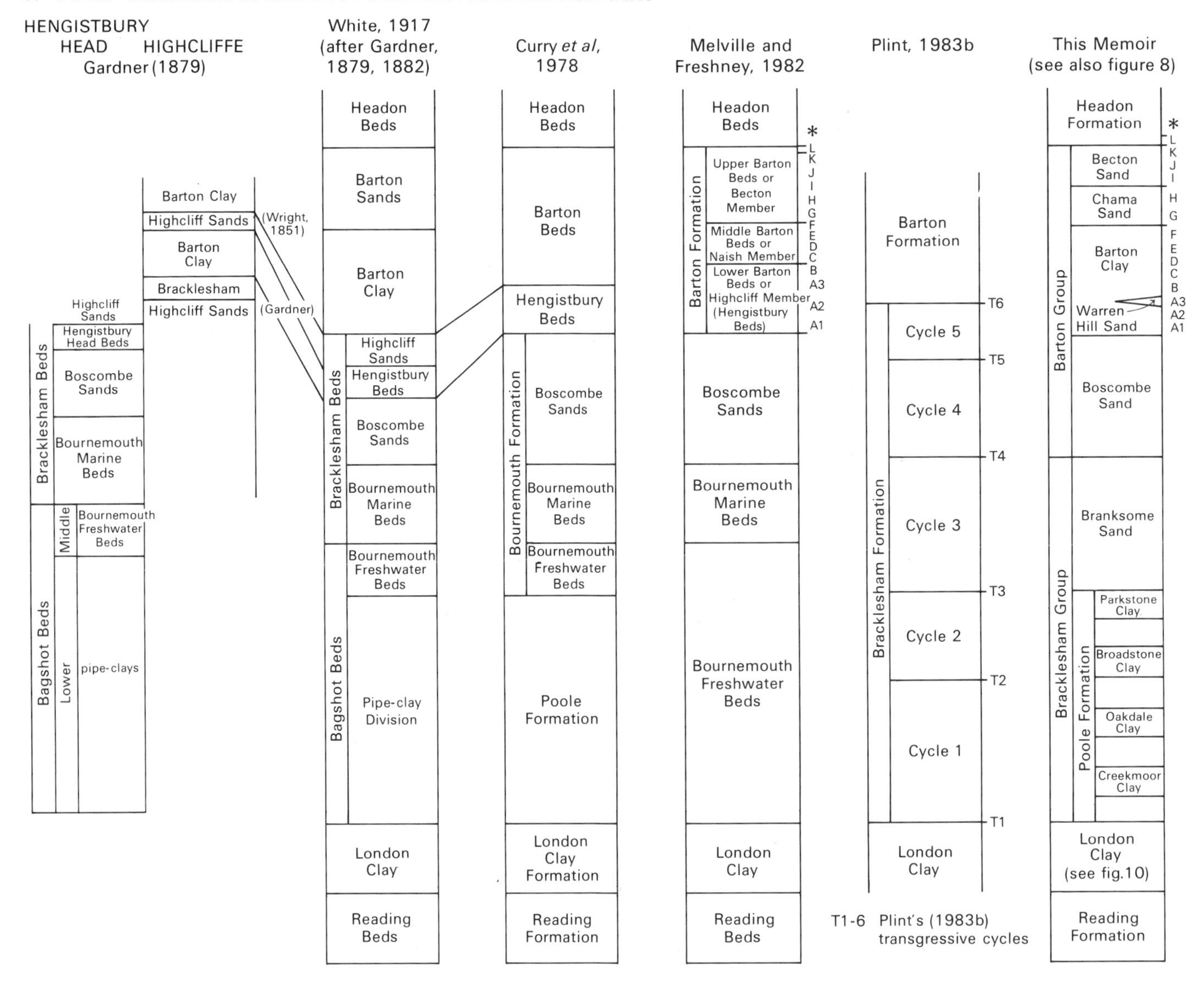

Figure 9 Selected stages in the evolution of the Palaeogene rocks of the district. (Implied correlation is by relative position in vertical column; actual correlation where different is by tie lines).

where it consists predominantly of red-mottled clay with sporadic beds of sand and lenticular pebble beds (Edwards and Freshney, 1987b).

Red beds are only locally present in the Wytch Farm boreholes, in the strata classified as Reading Formation by the oil company geologists. A reinterpretation of the gamma-ray logs of these boreholes (Figure 10), aided by palaeontological reports on the ECC Ball Clays Ltd (hereafter referred to as ECC) Rempstone Borehole [SY 9858 8507] (Figure 10), and by a comparison with the fuller Reading Formation succession in Sandhills No.1 Borehole, suggests that the 4.5 to 11 m-thick sand at the top of the so-called 'Reading Formation' at Wytch Farm (Colter and Havard, 1981) and Rempstone corresponds to the Tilehurst Member (basal London Clay). The Reading Formation is represented by the underlying 6 to 23 m of patchily red mottled silty clay and clayey sand (Bristow and Freshney, 1987, fig.1). In the Rempstone Borehole, the Reading Formation commences with a 2.5 m-thick bed of glauconitic sand with flints, above Chalk, succeeded by 9.5 m of dark to light grey-brown clayey sand.

In the gamma-ray logs of the Wytch Farm boreholes, a 1.5 m-thick bed of sand is present in the middle of the Reading Formation (Figure 10). This sand may be present at Rempstone, but was not recognised due to core loss. In the Wytch Farm No. X14 Borehole [SY 9804 8526], reddened strata occur between the Tilehurst Member and the above-mentioned sand, whilst in the Wytch Farm A-series boreholes, reddened beds occur below the 1.5m-thick sand. Red beds were not recorded in any of the other boreholes.

ECF, CRB

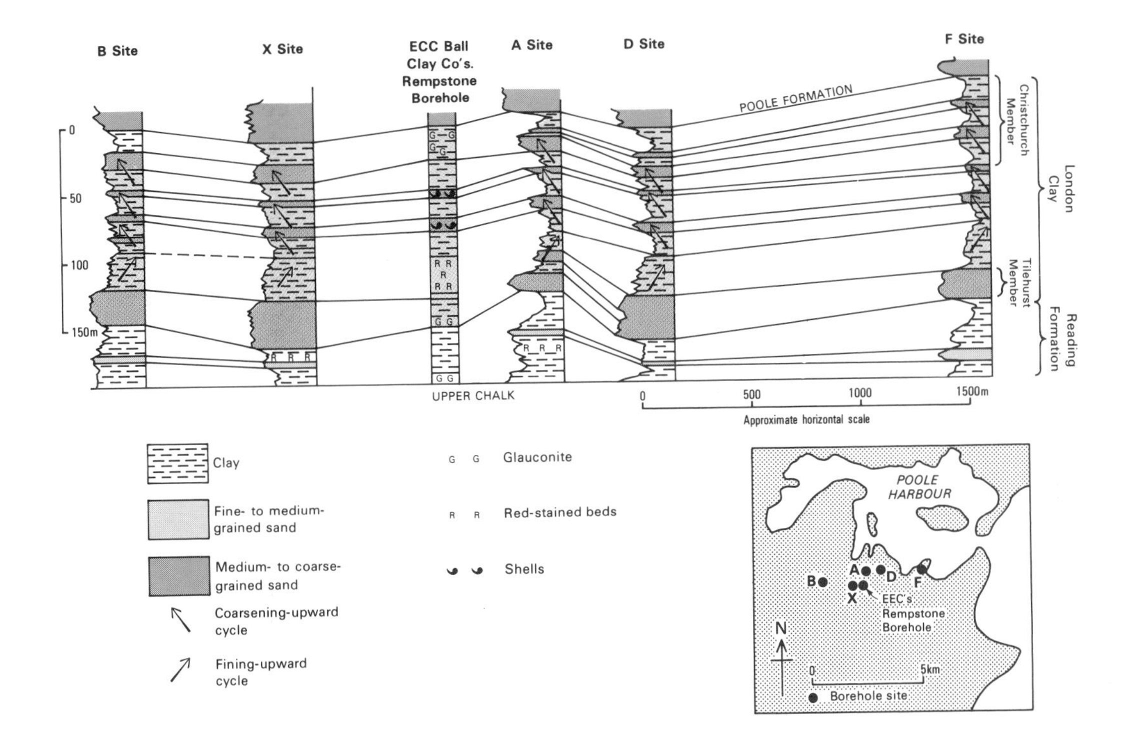

Figure 10 Correlation of the Reading Formation and the London Clay in boreholes in the Wytch Farm area.

LONDON CLAY

Stratigraphy

The outcrop of the London Clay extends over a 2 to 4 km-wide tract in the north-west of the district between Lytchett Minster [SY 96 93] and Holt [SU 03 04]. There is an extensive subdrift outcrop along the River Stour as far as Longham [SZ 07 98], and the formation occurs at depth over the whole of the district.

Because of the north-westerly overstep of the Poole Formation, and also an eastward facies change from fluviatile and lagoonal sediments at the base of the Poole Formation (Creekmoor Clay) into intertidal sediments at the top of the London Clay (Christchurch Member) in the subcrop to the east, the unexposed London Clay is thicker than at outcrop. The London Clay subcrop in the south-west of the district, in the Wytch Farm and Rempstone boreholes, commences with the Tilehurst Member (p.22), and thus includes strata formerly classified as the Reading Formation (Colter and Havard, 1981). King (1981) included the Tilehurst Member within the Oldhaven Formation beneath the London Clay, but Edwards and Freshney (1987a) regard it as part of the London Clay. The sequence comprises 4.5 to 11 m of glauconitic, silty, fine-grained sand with a sharply defined top and bottom.

In the Wytch Farm boreholes, the Tilehurst Member is succeeded by a fining-upward sequence, followed by three, and in places four, coarsening-upward sequences with a total thickness of between 32 and 38 m (Figure 10), but it has not been possible to correlate these with certainty with the sequences recognised at outcrop. Farther east, in the deeper part of the basin, five coarsening-upward cycles occur. Ideally, each commences with a thin pebble bed, or a glauconitic bed, succeeded by clay, which passes up into fine-grained sand, or into laminated/interbedded sand and clay laid down at the close of each cycle.

The reddened basal clay member and the two overlying sand members at outcrop in the north-west have been named the West Park Farm Member, the Warmwell Farm Sand and the Lytchett Matravers Sand respectively. The West Park Farm Member is probably the red- stained equivalent of the fining-upward cycle and the first of the coarsening-upward cycles recognised at Wytch Farm. In the tract extending eastwards from Wimborne to Longham [SZ 06 98], the London Clay above the West Park Farm Member is not divisible into mappable members.

The thickness of London Clay above the West Park Farm Member is about 20 m at outcrop south of the River Stour, but there is some local variation (as thin as 12 m in the Oakley to Ashington [SZ 005 983] area, and up to 50 m north of the Stour), due to overstep by the Poole Formation. Near Lytchett Minster, this part of the sequence is about 30 m thick, and in the Beacon Hill Borehole [SY 9761 9446], where the younger Christchurch Member is present, it is 40 m (Bristow and Freshney, 1986a). At Wytch Farm the thickness varies from 32 to 38 m. ECF, CRB

West Park Farm Member

There are three main reasons for regarding the reddened clays of the West Park Farm Member as part of the London Clay, and not the Reading Formation: i) in places, reddened clay, which does not show the high gamma-ray counts normally recorded in the Reading Formation, overlies blue-grey silty clay typical of the London Clay, ii) gamma-ray logs of the reddened strata show coarsening-upward cycles characteristic of the London Clay, and iii) there is at least one bed of well-rounded flint pebbles within the reddened clays (Figure 11).

At the base of the West Park Farm Member there is a distinctive 1 m-thick bed of glauconitic sand or sandy clay, commonly shelly, with glauconite-coated flint pebbles and cobbles. Near Kingdown Farm [ST 97 04], it forms a gravel-like pavement or a gravelly orange and grey sandy clay on the field surfaces. The pebbles consist dominantly of well-rounded, small flints, (generally up to 4 cm across), but larger scattered, well-rounded and angular flints and blocks of coarse-grained sandstone up to 10 cm across also occur. North of Wimborne, a glauconitic sandy clay, up to 4 m thick, crops out over a small area and can be mapped separately, as, for example, north-west of High Hall [ST 996 034]. Over much of the district, however, the glauconitic bed probably does not exceed 0.3 m in thickness and locally it is absent (p.27). At Wytch Farm, for example, red-stained, greenish grey, very silty clay, 7 m thick, directly overlies the Tilehurst Member.

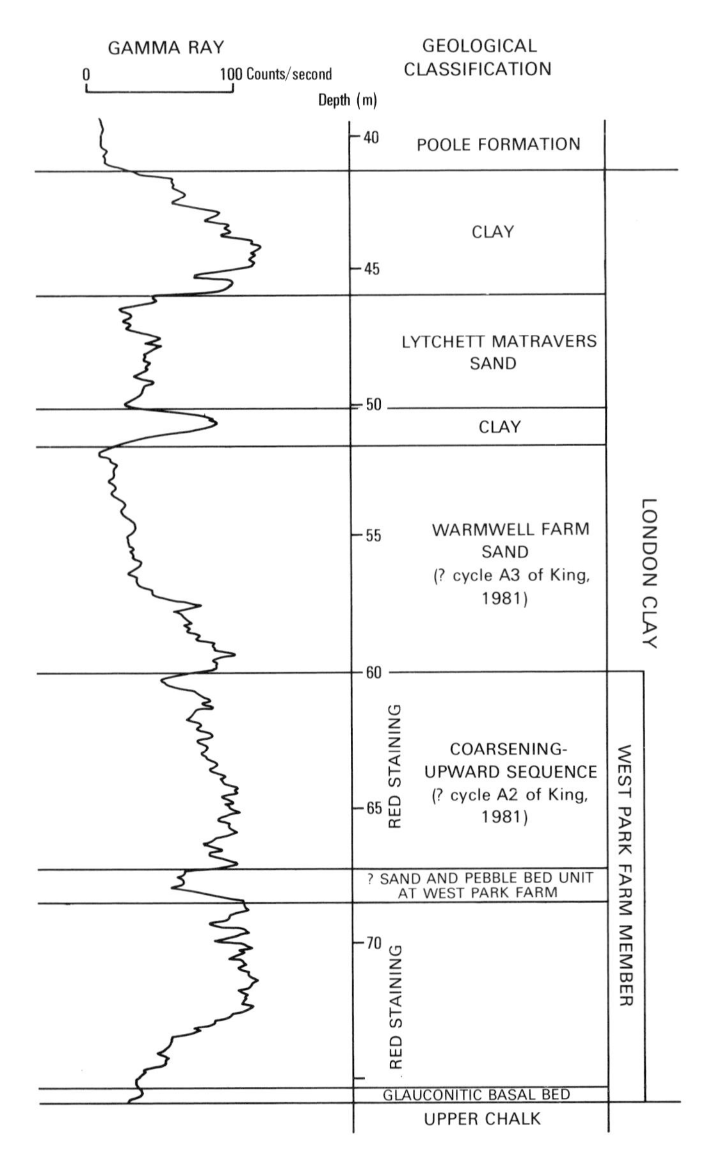

Figure 11 Gamma-ray log of the London Clay at Morden Heath (from data supplied by Pilkington Tiles Ltd).

The succeeding strata, some 8 to 30 m thick, consist principally of mottled orange and grey silty clay, though in many places, for example Knoll Manor Clay Pit [SY 973 977], the clay is strongly red or purplish red, and resembles the Reading Formation. There are also subordinate sands within the sequence. A sand and pebble bed, probably equivalent to the sand and pebble bed within the West Park Farm Member at the type locality [SY 9270 9644] west of the district (Figure 11), occurs within reddened clay north-west of Combe Almer [around 948 980], where it is thick enough to map over a 1 km tract. Thick beds of fine- to coarse-grained, locally cross-bedded, lignitic sand occur within the member in the west near Combe Almer [around SY 952 975] and in the north-west near Tadden [around ST 999 013]. Locally, such as south of Combe Almer, the whole of the West Park Farm Member is represented by sand. Near Kingston Lacy [around ST 973 012], sand at the base of the member occupies a steep-sided channel (White, 1917; Bristow, 1987a).

At two localities near Lytchett Heath [SY 964 945] in the south, and in the Beacon Hill Borehole, glauconitic grey sandy clay beneath the Warmwell Farm Sand is assumed to be the lateral equivalent of the West Park Farm Member.

ECF,CRB

Undivided London Clay above the West Park Farm Member

In the Wimborne to Longham area, where the interbedded sand members are absent or too thin to map, the London Clay above the West Park Farm Member is undivided.

Near Corfe Mullen, Wimborne and Canford Magna, a highly glauconitic, clayey, locally shelly, pebbly sand up to 2 m thick, the lateral equivalent of the Warmwell Farm Sand, rests on the West Park Farm Member, but is too thin to show on the maps. Elsewhere, in the Wimborne–Longham area, the bed immediately overlying the West Park Farm Member is commonly glauconitic and shelly. Near Canford Magna, impersistent beds of coarse-grained sand up to 6 m thick, and coarse-grained, glauconitic sandy clay at a higher level were encountered in several boreholes (Freshney et al., 1985). The log of a representative borehole in the undivided London Clay sequence is given on p.29. The thickness of the undivided sequence ranges from about 12 m in the Oakley to Ashington area [SZ 005 983] to about 30 or 35 m near Longham [SZ 065 980].

CRB

Warmwell Farm Sand

The member takes its name from the exposures in the lanes and banks near Warmwell Farm [SY 949 964]. It is generally up to 6 m thick, as in the type area, but locally reaches 15 m. It is well developed in the area south-west of Henbury Plantation [SY 966 980]. There, and south-west of Corfe Mullen, the member forms a prominent scarp up to 6 m-high, generally with powerful springs issuing from its base.

North of the River Stour, around Pamphill [ST 99 01], the Warmwell Farm Sand forms a 5 m-high scarp. East of the River Allen, the member persists, but does not form a prominent scarp.

The constituent sands, which are commonly well sorted, and symmetrical to negatively skewed, vary from very fine to fine grained, and are locally clayey. Pebble beds composed of well-rounded black flints occur at least at two levels. Most pebbles are between 1 and 2 cm across, but some reach 10 cm. The pebble beds vary from stringers of scattered pebbles to matrix-supported pebble beds up to 0.6 thick; most beds are less than 10 cm thick. CRB

Clay above the Warmwell Farm Sand

The Warmwell Farm Sand is succeeded by silty and sandy, bioturbated, olive-grey clay, interbedded with clayey fine- to very fine-grained silty sand. At the base of the clay, a breccia of tabular siltstone and fine-grained sandstone fragments with scattered rounded flints, up to 10 cm thick, is locally present (p.30). In places, mainly at the top of the succession, the interbedded sand becomes slightly coarser grained. The clay is generally 10 to 15 m thick, but is locally 25 m thick in the west, and up to 15 m thick in the north. The clay weathers to mottled orange and grey, but mottled red, yellow and grey clays occur locally. In the west, there are abundant 1 to 2 cm-thick, brittle, ferruginously cemented layers (?laterite). CRB

Lytchett Matravers Sand

The sand of this member forms extensive flat areas around Lytchett Matravers, where it is fine grained, commonly ferruginously cemented and locally hard enough to have been worked as building stone (White, 1917). It ceases to be a mappable unit, or is cut out beneath the unconformable Poole Formation, in the area extending from just north of Corfe Mullen to Wimborne, but reappears between Wimborne and the northern margin of the district. The maximum thickness is about 6 m south of the River Stour, and 8 m north of the Stour. No pebble bed has been noted in this sand south of the River Stour, but pebble beds are present in the Holt area to the north. Glauconite has been noted in the basal beds near Wimborne (White, 1917).

The Lytchett Matravers Sand commonly has a sharp feature break, locally associated with springs, at its lower boundary. CRB

Clay above the Lytchett Matravers Sand

The clay overlying the Lytchett Matravers Sand is similar to the clay above the Warmwell Farm Sand. It has been worked for the manufacture of bricks and pipes near Lytchett Matravers [SY 9485 9517]. It probably has a maximum thickness of 10 m around Lytchett Matravers, and 15 m near Holt, where it is overlain by sands of the Poole Formation. CRB

Christchurch Member

This member occurs only in the subcrop where it consists mainly of fine-grained, silty, commonly glauconitic and micaceous sand, interbedded with yellowish brown, often carbonaceous, colour-banded and laminated sand and silt. Claystone concretions and scattered black flint pebbles also occur in places. Palaeosols with or without rootlets are common. The type sequence is the succession between the depths of 191 and 247 m in the Christchurch Borehole [SZ 2002 9301]. It is also present in boreholes at Holdenhurst [SZ 132 953] (over 53 m), and at Knapp Mill, in Christchurch [SZ 1544 9380] (56.38 m). Comparable beds occur at Alum Bay in the Isle of Wight between Prestwich's (1846) Beds 7 and 13, and are probably present in the Beacon Hill Borehole [SY 9761 9446] between 28 m and 41 m. The member is 23 m thick in ECC Ball Clay Co's Rempstone Borehole, and geophysical logs of oil wells at Wytch Farm also suggest the presence of about 15 m of beds. Geophysical well log correlation between the Wytch Farm oil wells, Christchurch Borehole and BP Bransgore oil well [SZ 1958 0950], suggests that the intertidal deposits of the Christchurch Member pass laterally westwards into the dominantly fluviatile and lagoonal deposits of the basal Poole Formation (Creekmoor Clay and underlying sand). ECF

Biostratigraphy

Dinoflagellate cyst microfloras have been collected from 25 localities, foraminifera from eight localities, and molluscs at two localities of the London Clay (see Figure 2 and Table 1); they range in age from the *hyperacantha* to *varielongitudum* zones of Costa and Downie (1976) (Figure 8). These zones correspond to divisions A to C of King (1981, fig. 45).

The fossil evidence for dating the named members of the London Clay is sparse. Some 17 m of strata below the Warmwell Farm Sand in the Beacon Hill Borehole, the lateral equivalent of the West Park Farm Member, fall within the *hyperacantha* Zone (Division A1 of King). A few outcrop samples from the London Clay above the West Park Farm Member yielded microfloras possibly indicative of the *hyperacantha* Zone but, as discussed by Bristow and Freshney (1986b), they may be facies controlled, and not age-diagnostic.

In the Knoll Manor Clay Pit [SY 974 978] (Table 1, locality 76), 3 km west-south-west of Lambs Green, there is a rich molluscan fauna, associated with much glauconite and well-rounded black flint pebbles, in beds overlying the West Park Farm Member. The fauna is rich in numbers, but low in species (see p.29). On general considerations, the fauna is thought to indicate Division A, possibly A3 (*meckelfeldensis* dinoflagellate cyst Zone) (Wood, in Bristow and Freshney, 1986a).

In the Beacon Hill Borehole, strata thought to be the Warmwell Farm Sand yielded a *meckelfeldensis* microflora (Division A2 or A3).

Foraminifera and dinoflagellate cysts from several localities indicate Division B (i.e. *simile* Zone) for the clay above the Warmwell Farm Sand. Foraminifera from the Canford Magna No. 6 Borehole [SZ 0535 9789] (see p.29) suggest a stratigraphical level at about the Planktonic Datum of Wright (1972), within Division B (*simile* Zone) of King (1981).

In the Christchurch Borehole (Figure 2 and Table 1, locality 10) strata of *simile* Zone age have a thickness of at least 8 m. In the Beacon Hill Borehole (locality 42), 2.1 m of fine-grained sand towards the top of the London Clay, probably the Lytchett Matravers Sand, contain dinoflagellate cysts of *simile* Zone age.

At Beacon Hill, the topmost 4.5 m of the London Clay yielded *Wetzeliella lunaris,* probably indicative of the *varielongitudum* Zone. In Canford Magna No. 7 Borehole (Table 1, locality 57), where the London Clay is not divisible into members, the uppermost clay, sampled at a depth of 16 m, yielded *Wetzeliella lunaris,* again a possible indication of the *varielongitudum* Zone.

In the Christchurch Borehole, the *varielongitudum* Zone spans the 56 m of the Christchurch Member and about 21 m of the clay below. The zone is the youngest of the London Clay in this part of the Hampshire Basin. ECF, CRB

Conditions of deposition

Three out of the five coarsening-upward sequences (Divisions A to C) that King (1981) recognised throughout much of the Hampshire and London basins are present in the district. They represent the repeated progradation of a nearby shoreline or delta, followed by an abrupt marine transgression, commonly marked by a flint-pebble bed. The reddened strata of the West Park Farm Member, and the many probable laterite layers higher in the sequence, however, indicate periodic subaerial emergence. The coastline on the western margin of the basin probably trended north-east–south-west, and the provenance of heavy minerals indicates south-westward sediment flow along the coast (Morton, 1982).

The West Park Farm Member probably formed under conditions similar to those established by Buurman (1980) for the Reading Formation of the Isle of Wight. That is, in a marine or fluviomarine environment in which periods of sedimentation alternated with intervals of emergence and soil formation, in a climate with a fairly high annual temperature and a dry season.

The grain-size distribution of the Warmwell Farm Sand suggests that it was deposited on a shoreface or beach at the end of a regressive sequence (Divison A (?A3) of King, 1981). Similarly, the Lytchett Matravers Sand was probably deposited during regression at the end of a second cycle (Division B). North of Poole, in the area around Wimborne Minster, and probably also in the Lytchett Matravers area, shelly sediments of the third cycle (Division C, that is the clays above the Lytchett Matravers Sand) probably mark a coastline with a beach-barrier sand. Farther south, in Poole Harbour, Christchurch, and in Alum Bay in the Isle of Wight, sands of the younger Christchurch Member, which also fall within Division C, represent estuarine deposits at the mouth of an eastward-draining fluviatile system. The thicker beds of silty, fine-grained, glauconitic sand in this member were probably deposited in channels, and the laminated brown carbonaceous clays and fine-grained sands in interdistributary bays. Rootlets in some of the clays, and palaeosols without rootlets, indicate periodic subaerial conditions, possibly on a delta top.

The westward passage of the intertidal clays, sands and palaeosols of the Christchurch Member into the fluviatile and barrier sands, and back barrier lagoonal clays of the lowest part of the Poole Formation, indicates that the coastline was interrupted by an eastward-draining deltaic system that was constrained to the south by an active structure along the Isle of Purbeck Monocline. The delta distributaries debouched into the more fully marine area in the eastern part of the Hampshire Basin. ECF, CRB

Details

West Park Farm Member

At the northern end [SY 9589 9532] of the Lytchett Heath inlier, greyish brown, micaceous, fine-grained, sandy clay occurs in the bottom of a valley. At a second locality [SY 9644 9460], stiff, grey, red-stained clay was found, and at a third [SY 9647 9450], olive-grey, extremely silty clay with glauconite was augered. ECF

About 500 m west-north-west of Warmwell Farm, there is a thick development of dominantly medium-grained, but with some fine- and coarse-grained sand [SY 940 966], underlain locally by 0.6 m of sandy clay with a glauconitic base, resting on Chalk [SY 9401 9672]. At other places [SY 9420 9657; 9437 9664; 9440 9678], up to 0.6 m of mottled orange and grey clay rests on up to 0.3 m of glauconitic sand, in turn resting on Chalk.

At two localities [SY 9515 9772; 9536 9819] near Combe Almer, glauconite-coated pebbles from the basal bed are common on the surface. This basal bed is succeeded by up to 10 m of sand, locally [e.g. SY 9504 9803] very coarse grained, which in turn is overlain by yellowish brown sandy clay [SY 950 978; 9515 9808]. The basal glauconitic bed to the south-west of Combe Almer is overlain by some 2 to 4 m of fine-grained sand [around SY 9472 9712], which in turn is succeeded by dominantly red, or mottled red, yellow and grey clay. Where the lower sand is absent [e.g. SY 9490 9705], the glauconitic basal bed is succeeded by 2 to 5 m of mottled orange and grey silty clay, which in turn is overlain by coarse-grained sand. White (1917) noted that a ferruginously cemented basal sand was worked for building stone hereabouts.

White (1917, p.11) also noted a 6 m section in a pit [SY 953 970] near Combe Almer in strongly cross-bedded, fine- and coarse-grained sand. On the west side of the pit, a lenticle of laminated, pale grey silty clay, about 1.5 m thick, overlies the sand. East of Dullar Farm, sand with a glauconitic sandy clay base forms the basal bed of the member. A small pit [SY 9477 9749] low in the succession showed 1.4 m of fine-grained orange sand resting on 0.1 m of red clay, which in turn overlies more than 0.2 m of chalky clayey sand. In a chalk pit [SY 9445 9755] south-east of Dullar Farm, Reid (MS, BGS) saw 1.5 m of ferruginous sand and ironstone, resting on 0.1 m of greenish loam with rare green-coated flints, on Chalk.

Mottled orange and grey, silty and fine-grained sandy clay overlies the basal sand over most of the Dullar Farm spur; only rarely was reddened clay noted during the survey. The clay is succeeded by a sand, locally associated with a pebble bed [SY 9482 9807], which is the presumed equivalent of the sand and pebble bed in the middle of the West Park Farm Member at the type locality. Springs are common at its base.

In a pit [SY 9531 9753] west of Heron Grove, 0.6 m of red clay overlies 0.1 m of glauconitic clay, above Chalk. White (1917, p.11) noted the 'Bottom bed' resting on hardened Chalk in a pit [SY 9705 9735] just west of The Knoll. Higher strata consist of about 10 m of red and grey clay, and yellow and brown sandy clay, which in turn are overlain by up to 15 m of fine- to medium-grained, cross-bedded buff sand. Northwards, the clays thin and die out. The sand dies out abruptly east of Henbury [SY 9600 9825], but fine- to coarse-grained sand is present in the valley [SY 967 970] south-east of Henbury Plantation. There, a section [SY 9659 9684] exposes about 1 m of fine-grained sand with common small, well-rounded pebbles in the disturbed top 15 cm. Higher strata hereabouts consist of mottled red and grey, and orange and grey clay.

In the Knoll Manor Clay Pit [SY 974 978], more than 10 m of purple, red, and greyish green and red mottled clays underlie the lateral equivalent of the Warmwell Farm Sand; a similar section is seen in an old pit [SY 9890 9865] at Brog Street. North-east and north-west of The Knoll, red, and mottled red and grey clays are the dominant lithologies, and are visible in the old railway cutting [SY 978 983 to 9815 9815]. They were also augered in the sides of two old pits [SY 9640 9844; 9665 9840]. Mottled orange and grey silty clay and clayey silt occur east and south of The Knoll [SY 9818 9745; 9805 9700].

A borehole [SY 9937 9890] at the sewage works proved, beneath 5.18 m of drift, 1.07 m of stiff brown and blue, laminated, silty clay, overlying 0.76 m of greyish blue, silty fine-grained sand, which in turn rests on 1.22 m of brown and bluish grey, with some reddish brown, silty clay.

The West Park Farm Member was recognised in one of the Wimborne Bypass boreholes [SZ 0244 9909]; the log reads:

	Thickness m	*Depth* m
London Clay		
Clay, very silty, with pockets of sand, and some gravel and fossil fragments; stiff, laminated, dark grey with occasional green specks	0.8	4.6
WEST PARK FARM MEMBER		
Clay, silty, mottled brown/pale grey, stiff, fissured	0.7	5.3
Clay, silty, with siltstone, mottled reddish brown and grey, very stiff	0.5	6.8
Clay, silty, with siltstone, mottled reddish brown/pale brown and grey, hard	9.7	16.5
Clay, very sandy, passing into a clayey, fine-grained sand; pale grey, stiff	1.0	17.5
Clay, silty, bedded, fissured, grey, becoming brown	1.5	19.0
Clay, silty, mottled yellow-brown/grey, very stiff, fissured	1.0	20.0

Wells on the outskirts of Wimborne [SZ 0222 9932], and at Canford Magna [SZ 0323 9892], proved the thickness of the member to be 19.8 and 12.2 m respectively. An incomplete thickness of 17 m was recorded in a well near Merley Hall [SZ 0057 9905].

Some 8.8 m of greenish glauconitic sands, with a basal 0.3 m-thick pebble bed, resting on Chalk, were recognised in the borehole [SZ 0323 9892] at Canford Magna. Elsewhere [e.g. SZ 0104 9996], the glauconitic sands and pebbly basal bed are absent and the basal 5 m consists of stiff, dark greyish blue, fissured silty clay, in part laminated, and patchily buff-stained. Up to 1.3 m of silty, fine-grained sand occurs within the clay in some of the Wimborne Bypass boreholes.

Some 6 m of red-mottled clay, just below the base of the Warmwell Farm Sand, were formerly exposed in a pit [SU 0063 0073] north-east of Wimborne cemetery.

The detailed successions of several boreholes near the Wimborne Waterworks vary, but generally there is an upper unit of clay, or clay and sand, up to 13 m thick, above 4 to 12 m of sand (Whitaker and Edwards, 1926). Some 600 m south-east of the Waterworks, a borehole proved the following descending sequence [SU0115 0061]; red-mottled clay, 13.4 m; dark clay (or clay and sand), 6.1 m; dark clay with pyrites, 1.8 m; sand with flint pebbles, containing black angular flints at the bottom, 1.2 m, on Chalk (Whitaker and Edwards, 1926). In another borehole [SU 0112 0024] farther south, beneath 3.3 m of made ground and alluvial gravel, the following sequence was found:

	Thickness m	*Depth* m
WEST PARK FARM MEMBER		
Clay, mottled grey and red, firm to stiff	1.5	4.8
Clay, grey, mottled red, silty, very stiff, with layers of grey clayey silt	3.0	7.8
Clay, red, mottled bluish grey, silty, very stiff		
Clay, mottled brown and bluish grey, patchily	1.8	9.6
red, silty, fissured, very stiff	2.5	12.1
Clay, silty, grey, fissured, very stiff	4.7	16.8
Sand, medium-grained, grey, with fine gravel	1.7	18.5
Sand, medium-grained, grey	0.8	19.3
Upper Chalk		
Chalk, white, rubbly to blocky	1.5	20.8

A similar sequence was proved in a nearby borehole [SU 0108 0002] (Bristow, 1987b).

East of the River Allen, between the waterworks and Clapgate [SU 010 024], the member is 24.4 m thick [SU 0093 0235]. In this area, mottled orange and grey silty clay occurs, but mottled red and grey clay was found at one place [SU 0104 0237] at the top of the member. Farther north, a basal glauconitic sandy clay or a clayey glauconitic sand crops out over a 300 m-long tract [ST 997 034] north-west of High Hall. White (1917) recorded green-coated flints and fragments of oyster shell in the fields hereabouts. Irregular pipes in Chalk, filled with coarse-grained loamy sand, with flint pebbles and angular flints, were seen at the top of a pit [ST 9980 0323] by Reid (MS map, BGS).

The 'greenish Bottom bed' was formerly exposed in the side [ST 9981 0300] of the lane north-west of High Hall (White, 1917). Farther south, glauconitic clayey sand was augered in the lane bank [ST 9843 0038] close to a site [ST 9840 0037] where Reid (MS. map, BGS) saw rounded black flint pebbles and angular flints resting on Chalk.

Glauconitic clay and sand, up to 0.9 m thick and associated with well-rounded flint pebbles, occur at the base of the outlier [ST 9700 0025] east of Barford Farm, and are overlain by red, and mottled red and yellow clay. Weakly glauconitic clay was found at one place [ST 9627 0300] on Badbury Rings, where it is succeeded by 2 to 3 m of mottled orange and grey, locally red-stained clay and capped by a pebble bed which is composed dominantly of well-rounded flints. Glauconitic clay was proved at two places only [ST 9685 0314 and 9681 0305], close to the base of the outlier of mottled orange and grey, locally red-stained, silty clay east of Badbury Rings. On the smaller outlier [ST 9685 0330] north-east of the Rings, glauconitic mottled orange and grey clay is associated with rounded flint pebbles.

An extensive outcrop of a non-glauconitic pebble bed (see p.24) caps the Chalk ridge that curves around the west, north and north-east of King Down Farm [ST 965 036 to 979 041]. At only one place [ST 9681 0402] were glauconitic sandy clay and glauconite-coated flints noted. The chalk surface of the field [ST 9795 0395] south-east of the King Down Farm outlier is covered with a thin layer of pebbles.

Fine- to coarse-grained, poorly sorted sand with thin clay seams forms the basal bed of the West Park Farm Member from Coneygar Copse in Kingston Lacy Park, south-eastwards towards Cowgrove, and eastwards towards Hound Hill. A section [ST 9722 0138] in Coneygar Copse appears to occupy a steep-sided channel (White, 1917):

	Thickness m
Soil; thin, sandy.	
WEST PARK FARM MEMBER	
Brown loamy sand, with inclusions of finer greyish brown loam	0.61 to 2.13

Roughly laminated greyish brown sandy clay, thinning out eastwards.	0.15
Coarse brown loamy sand of flint and quartz, full of rounded and angular bits of brown clay and ironstone, small angular chips of flint, and a few unworn flints. The materials are loosely cemented in places by iron-oxide, and the structure is 'confusedly' current-bedded	0.91
Coarse, black-stained flint and quartz sand	0.61
Light brown and reddish sand with beds of soft sandstone, not well shown	?0.91
Yellow and brown sand, becoming very coarse in the lower part, where it contains many small angular pieces of flint	1.83

White (1917) noted that the junction of the 'Reading Beds' and Chalk was almost vertical in a temporary exposure just south of the pit. There, the basal bed consists of sand with a 'selvage of stiff grey loam, containing pellets of chalk and small concretions of ironstone, and exhibiting a brecciate appearance on bared surfaces. The characteristic pebbles, unworn flints, and perforations of the Reading Bottom-bed are wanting.' The near-vertical contact, and the absence of a glauconitic basal bed, suggested a faulted junction to White (1917). Reid (in White, 1917), however, thought that the steep contact was a channel margin. The latter interpretation is probably correct. Unknown to either author, the sand in the pit is not part of an outlier, but continues eastwards as a narrow cap to the east-south-east-trending ridge along which the boundary falls gradually east-south-eastwards. The contact of the West Park Farm Member and the Chalk appears to be normal along this tract. For the most part, coarse-grained, orange-brown sand occurs along this ridge, but locally [ST 9773 0118; 9780 0109; 9790 0104] grey clay, associated with fine-grained sand, was found.

East of the River Allen, weakly glauconitic greyish brown clay was found at one locality [SU 0076 0377] near Stanbridge, and glauconitic sandy clay at another [SU 0080 0378]. An extended auger hole [SU 0099 0393] 250 m north-east of the last place, proved:

	Thickness m	*Depth* m
Topsoil	0.4	0.4
WEST PARK FARM MEMBER		
Clay, silty, mottled orange and grey	0.2	0.6
Clay, grey, red-stained	0.8	1.4
Clay, creamy	0.3	1.7
Clay, red-stained	0.2	1.9
Clay, creamy and ?chalky	0.2	2.1
Clay, sandy, greenish grey, with fragments of chalk, thick-shelled bivalves and patinated flints	1.7	3.8
Sand, clayey, glauconitic	0.2	4.0

North-west of this occurrence at Stanbridge, glauconitic sandy clay was augered around the tops of old chalk pits [SU 0068 0396; 0080 0404; 0086 0415]. Very glauconitic sandy clay and clayey sand was found at one place [SU 006 040] about 4 m above the base of the West Park Farm Member. About 5 to 6 m above the base at a nearby locality [SU 0058 0402], glauconitic clayey sand was augered above weakly glauconitic, greyish brown clay. An extended auger hole [SU 0102 0391] 600 m east-south-east of Stanbridge church proved the following succession:

	Thickness m	*Depth* m
WARMWELL FARM SAND (OR ?HEAD)		
Sand, coarse-grained, buff, wet	1.0	1.0
WEST PARK FARM MEMBER		
Clay, dark greyish brown	0.1	1.1
Clay, very stiff, mottled red/grey	1.3	2.4
Clay, silty, very stiff, dark grey, with ?siderite spherules	0.3	2.7
Clay, red-stained	1.1	3.8

A borehole [SU 0240 0413] downdip near Holt proved, beneath the higher members of the London Clay, 10.7 m of mottled clay, resting on 1.5 m of pale green sand, which in turn rests on Chalk. Almost 1 km to the south, another borehole [SU 0231 0318] penetrated a total thickness of 16.8 m of the member, which consists of red and mottled clay with a sand bed that was, according to the drillers, 'brilliant green' near the base. CRB

UNDIVIDED LONDON CLAY ABOVE THE WEST PARK FARM MEMBER

A section in the Knoll Manor clay pit [SY 974 978] was recorded by Cooper, Hooker and Ward (1976), who made a collection of fossils. During the recent survey, three sections were measured in this pit; the first [SY 9742 9770] showed:

	Thickness m
London Clay	
Sand, very fine-grained, buff, interbedded with grey clay; sand in beds up to 0.4 m thick is dominant; clays are thinner and up to 12 cm thick, but with fine-grained sand laminae; thin lateritic layers	2.0
Sand, very clayey, laminated	1.0

The second section [SY 9739 9767] exposed:

	Thickness m
Head	
Clay, gravelly	0.6
London Clay	
Clay, greyish brown, with fine-grained sand partings	0.4
Sand, fine-grained, buff	0.3
Clay, grey, with fine-grained sand partings	1.0
Sand, fine-grained, with thin (up to 10 cm) clay partings	1.0
Sand, fine-grained, clayey, glauconitic	0.6
Unexposed	c.0.6
WEST PARK FARM MEMBER	
Clay, mottled red and grey	2.0

The third section [SY 9736 9766] revealed:

	Thickness m
London Clay	
Clay and fine-grained sand (poorly exposed)	c.4.0
Sand, clayey, highly glauconitic	0.6
Pebble bed of small (generally less than 2 cm) well-rounded flints; shelly, with common *Turritella* and bivalves; locally ferruginously cemented; spring at base	0.1
WEST PARK FARM MEMBER	
Clay, mottled red and grey	2.0

The macrofauna, identified by Mr C J Wood, from the shelly pebble bed includes: *Ditrupa plana, Rotularia bognoriensis, ?Ancistosyrinx* sp., *Euspira glaucinoides, Turritella* cf. *interposita, Caestocorbula* sp?, *Callista (Microcallista) proxima, Corbula* sp?, *Dosiniopsis bellovacina, Glycymeris brevirostris, Nemocardium plumstedianum, Nucula* sp., *Orthocardium* cf. *subporulosum* and *Striatolamia macrota.* Although not diagnostic, the fauna suggests Division A, possibly Division A3, of King (1981). A sample from the glauconitic clayey sand in the third section yielded one example of the facies-controlled dinoflagellate cyst *Apectodinium* sp., indicative of a nearshore environment of deposition.

In an old brick pit [SY 9893 9864] north-east of Brog Street, and in a nearby ditch [SY 9899 9873], about 0.6 m of glauconitic sand rests on mottled red and grey clay of the West Park Farm Member.

This glauconitic sandy bed crops out north-west of Merley House, and was penetrated in two piston-sampler boreholes [SZ 0055 9870; 0050 9887]. In the first, the basal 1 to 1.5 m were highly glauconitic and bright green, and yielded the dinoflagellate cysts *Apectodinium paniculatum, A. quinquelatum, Cordosphaeridium gracile* and *Homotryblium tenuispinosum,* indicative of an inner neritic environment of deposition and of the *hyperacantha* Zone or a younger zone. Scattered glauconite grains were present some 2.2 m above the base in the second borehole. In a borehole [SZ 0226 9899] farther east at Oakley, 0.35 m of firm to stiff, thinly laminated, grey silty clay with dark green inclusions and fine-grained sand and shell fragments, overlies the West Park Farm Member. In another nearby borehole [SZ 0224 9900], no glauconite was seen in the basal 0.6 m of grey sandy silt and laminated silty clay. In a well [SZ 0222 9932] north of the river, the 1.83 m of beds above the West Park Farm Member are described by the driller as 'grey-green sandy clay'.

Higher strata were encountered in many of the boreholes for the Wimborne By-pass. A typical borehole [SZ 0171 9868] proved the following, beneath 3.8 m of river terrace deposits:

	Thickness m	*Depth* m
London Clay		
Sand, fine-grained, silty, mottled orange and light brown (0.20 m proved)	0.2	4.0
Silt, clayey, fine-grained, sandy, firm, grey	1.0	5.0
Clay, silty, thinly bedded, grey, firm to very stiff, with scattered siltstones	1.8	6.8
Silt, clayey and fine-grained sandy, grey	2.2	9.0
Clay, silty, fine-grained sandy, thinly laminated, grey, very stiff, with siltstone	1.7	10.7
Silt, fine-grained sandy, grey	0.3	11.0
Silt, clayey, firm to stiff, with fine-grained sand laminae,	1.0	12.0
Sand, fine-grained, silty	2.0	14.0
Sand, very fine-grained, silty, with 0.1 m bands of soft, thinly laminated, pale grey, very silty clay	4.2	18.2
Silt, clayey, fine-grained sandy, with fine-grained sand laminae; shelly, very stiff, fissured	2.3	20.5

Dinoflagellate cysts from grey, fine-grained, sandy clay in two boreholes [SZ 0139 9792; 0098 9823] near Merley House belong to the *homomorpha* plexus and indicate a nearshore marine environment; the presence of *Wetzeliella meckelfeldensis* indicates the *meckelfeldensis* or a younger zone. Additional dinoflagellate cysts from the second borehole include *Homotryblium tenuispinosum*, indicative of a lagoonal environment.

At Oakley [SZ 0207 9870], 14.6 m of bluish grey, fine-grained sandy clay were encountered beneath 7 m of river terrace deposits and the Poole Formation. The foraminifera *Anomalinoides nobilis, Cibicidoides alleni* and *Karreriella* cf. *danica* were found in the uppermost 3 m of the London Clay. The Middle Eocene dinoflagellate cyst *Heteraulacacysta leptalea* was recovered from a sample at a depth of 16.5 m.

The London Clay in a borehole [SZ 0447 9965] near Little Canford consists of 7 m of dominantly olive-grey, shelly, silty clay, resting on 2 m of very sandy clay, which rests on 8.5 m of chocolate brown clay with thin interbeds of fine-grained sand. A sample 4 m from the top of the sequence yielded the foraminifera *Alabamina westraliensis, A. nobilis, Brizalina pulchra, Cibicidoides. alleni, Globigerina* gr. *linaperta, Gyroidinoides octocameratus, Nonion laeve, Pullenia quinqueloba, Pulsiphonina prima* and *Spiroplectammina adamsi.* A sample from near the bottom of the borehole yielded the long-ranging species *Nonion graniferum, N. laeve, Pararotalia curryi* and *P.* sp., indicative of shallow-water, low salinity conditions. The poor dinoflagellate cyst flora includes members of the *homomorpha* plexus; the presence of *Wetzeliella meckelfeldensis* 6 m from the top indicates the *meckelfeldensis* or a younger zone. A borehole [SZ 0400 9908] 750 m south-west of the above proved 17.1 m of clay with a 6 m-thick bed of coarse-grained sand near the top; some of the underlying clay contains much coarse-grained patchily glauconitic sand. Dinoflagellate cysts include members of the *homomorpha* plexus; scattered *Lingula* sp. were present near the base of the sequence.

Beds of coarse-grained sand and coarsely sandy clay up to 2.6 m thick were proved farther south, in boreholes [SZ 0406 9859; 0425 9797] east and south-east of Canford Park. The first contained much glauconite in the lowest 8.8 m. The dinoflagellate cysts *Apectodinium quinquelatum, Adnatospaeridum multispinosum, Ceratiopsis wardenensis* and *Cordospaeridium inodes* were recorded in both boreholes, and *Melitasphaeridium asterium* in the second. Foraminifera from the second borehole include *Pararotalia* cf. *globigeriniformis* and *Nonion laeve,* indicative of an inshore marine environment.

In a borehole [SZ 0439 9769] 500 m west of Knighton, the top part of the London Clay, beneath 9.3 m of river terrace deposits and Poole Formation sands, consists of 7.3 m of very clayey coarse-grained sand passing down into coarsely sandy clay. These latter beds rest on 2.9 m of medium grey, silty clay, which in turn rest on 0.9 m of coarsely sandy clay. The dinoflagellate cysts *Homotryblium abbreviatum* and *H. tenuispinosum, indicative of the homomorpha* plexus, and *Dracodinium similis* were recovered from the 2.9 m of medium grey, silty clay, together with the foraminifera *Ammodiscus siliceus, Dorothia* cf. *fallax, Haplophragmoides* cf. *walteri, Spiroplectammina adamsi* and *Trochammina* cf. *globigeriniformis.* The uppermost 3.6 m of the London Clay in borehole [SZ 0443 9728] 550 m south-west of Knighton, consisting of medium grey, finely sandy clay, also yielded *Dracodinium similis.*

The 20.4 m of beds in borehole [SZ 0535 9789] 400 m north-east of Knighton, consist of finely sandy clay, which is shelly and glauconitic in places. Dinoflagellate cysts of the *homomorpha* plexus were recorded. The foraminifera from 13 to 18 m depth suggest shallow-water shelf conditions; they include *Acarinina pentacamerata, Alabamina westraliensis, Anomalinoides noblis, Asterigerina* cf. *aberystwythi, Brizalina anglica, Cibicides* cf. *lobatulus, C. proprius, C.* cf. *pygmeus, Dorothia* cf. *fallax, Elphidium hiltermanni, Guttulina problema, Haplophragmoides* cf. *walteri, Karreria fallax, Lenticulina inornata, Melonis affinis, ?Neoponides* sp., *?Nonion* sp., *Pullenia quinqueloba, Pulsiphonina prima, Quinqueloculina reicheli, Spiroplectammina adamsi, Uvigerina batjesi* and *Voorthuysenella* sp. By comparison with the Bunker's Hill Borehole [SZ 3038 1498] (Edwards and Freshney, 1987b), the fauna suggests a stratigraphical level at about the 'Planktonic Datum' of Wright (1972), within Division B of King (1981).

Some 14.9 m of fine-grained, very sandy clay and clayey fine-grained sand in a borehole [SZ 0578 9696] 1.1 km south-east of Knighton yielded the dinoflagellate cysts *Homotryblium abbreviatum, H. tenuispinosum* and *Wetzeliella lunaris*, the last of which may indicate the *varielongituda* Zone, which would be the youngest dinoflagellate zone recognised in the London Clay of the Stour Valley.

West of Longham, a borehole [SZ 0593 9882] proved the following London Clay succession beneath 3.1 m of drift: yellowish brown, sandy clay, to 8 m; olive-grey, stiff, sandy clay, to 16 m; on glauconitic, very sandy clay to clayey,fine-grained sand, with a layer of black flint pebbles at 18 m, to 20 m. CRB

WARMWELL FARM SAND AND OVERLYING CLAY

The base of the Warmwell Farm Sand is clearly defined by springs and a sharp feature at the base of a 5 to 6 m-high escarpment north-west of Lytchett Matravers.

Sections around [SY 9470 9662] in the road bank 350 m north-west of Warmwell Farm revealed:

	Thickness m
London Clay	
Clay, very silty, mottled orange and grey, and fine-grained sandy clay	1.8
WARMWELL FARM SAND	
Sand, fine-grained, with a lateritic top; bed of scattered small (up to 2.5 cm) rounded flint pebbles in middle of unit	0.8
Clay, sandy, coarse-grained, passing up into a clayey fine-grained sand; pebble bed (pebbles up to 8 cm across) 0.3m above base of section	1.0

About 350 m to the north-east, an old quarry [SY 950 968] exposes 1.8 m of horizontally bedded, fine-grained sand. White (1917, pp.17–18) recorded three similar sections at this and nearby localities [SY 950 968; 9547 9662; 9546 9672]. The third one showed:

	Thickness m
London Clay	
Clay, sandy, mottled greyish brown, with clay-ironstone nodules	1.5
WARMWELL FARM SAND	
Flint pebble bed	0.1
Sand, fine-grained, with bands of grey laminated loam; thin seam of ironstone in the upper part	2.1
Bed of flint and small quartz pebbles in a coarse-grained sand matrix	0.2
Sand, fine-grained, buff and white, cross-bedded, with fine seams and specks of lignite near the top	2.4

White (1917) regarded the lower sand in the above sections as part of the 'Reading Beds'. However, the sands are here considered to form an indivisible unit, the Warmwell Farm Sand.

At one place [SY 962 982], near Henbury, the Warmwell Farm Sand is represented by a glauconitic sandy clay. The overlying beds consist of up to 25 m of mottled orange and grey, silty clay, which is locally lateritic.

South of Warmwell Farm, fine-grained yellow sand and pebbles are thrown out of a badger sett [SY 9504 9616]. An old pit [SY 9519 9617] 150 m to the east, dug into yellow fine-grained sand, is capped by mottled orange and grey silty clay. Some 130 m south-east of this, fine-grained sand, with some well-rounded flint pebbles, is exposed. Sections [SY 9545 9612; 9549 9614] a further 150 m to the east provide the following composite sequence:

	Thickness m
London Clay	
Clay, silty, mottled orange and grey, lateritic; scattered matrix-supported, well-rounded flints, and tabular and angular ferruginous siltstone and fine-grained sandstone clasts in basal 10 cm; most flints are less than 1 cm diameter, but some are up to 3 cm, and one 10 cm diameter	0.6
WARMWELL FARM SAND	
Sand, very fine-grained, mottled orange and grey	0.1
Sand, fine-grained, yellow and orange (poorly exposed)	c.1.0

Another section [SY 9550 9614] 10 m to the east revealed:

	Thickness m
LONDON CLAY	
Clay, sandy, mottled orange and grey, with scattered well-rounded flint pebbles (up to 15 mm diameter) in the basal 5 cm	0.6
Sand, very fine-grained, mottled orange and grey	0.1
Sand, very fine-grained, buff	0.1
Breccia of tabular and angular ferruginous siltstone and fine-grained sandstone fragments; one rounded flint pebble	0.1
WARMWELL FARM SAND	
Sand, fine-grained, buff-brown	0.2

Mottled orange and grey silty clay above the Warmwell Farm Sand was worked for brickmaking in pits [SY947 956; 959 957; 9582 9623] near Lytchett Matravers. South-west of Lytchett Matravers, the sands, locally pebbly [SY 9410 9507], form a prominent scarp about 5 m high. The overlying clay is about 5 m thick.

A section [SY 9602 9538] in a gully 1.3 km east of Lytchett Matravers shows:

	Thickness m
WARMWELL FARM SAND	
Sand, fine-grained, orange, horizontally bedded	2.00
Pebble bed of very well-rounded flints 0.5 to 3 cm in diameter, matrix-supported in a coarse-grained sand	0.06
Sand, clayey, coarse-grained	0.15
Sand, fine-grained, buff-orange	0.35

CRB

A ditch [SY 9645 9397] near Lytchett Minster showed 2 m of orange to buff, very silty clay containing claystone nodules and a stringer of black rounded flint pebbles. Another ditch [SY 9641 9382] showed yellowish brown, extremely silty clay grading into clayey, fine-grained sand with several laterite layers. Farther east-south-east [SY 9719 9365], yellowish grey, clayey, very fine-grained sands were seen in a third ditch. West of Lytchett Minster around [SY 957 930], much ferruginously cemented clay and clayey fine-grained sand debris occurs in the soil. ECF

North of Crumpet's Valley Farm, the Warmwell Farm Sand is about 15 m thick and the overlying clay up to 10 m thick. A section [SY 9645 9702] in an old pit reveals:

	Thickness m
London Clay	
Clay, silty, mottled orange and grey; a clay breccia at the base, up to 0.1 m thick, with one rounded flint pebble and a clay matrix.	
WARMWELL FARM SAND	
Sand, very fine-grained, buff-orange, with thin clay layers and an irregular base	0 to 0.8
Sand, fine-grained, clayey, thinly bedded; an	

impersistent pebble bed about 0.5 m above the base; a lateritic layer, 1 to 2 cm thick, between 0.1 and 0.25 m above the base; irregular clay bed, 1 to 2 cm thick above the laterite.	c.1.0
Sand, fine-grained, buff, locally lateritic; 1 m exposed, but probably continues to bottom of pit, 2 m lower.	3.0

Locally, there are lateritic nodules at the base of the clay bed, with moulds of *Turritella* and the bivalve *?Dosiniopsis bellovacina*. The uppermost sand bed is either channelled into, or is strongly affected by dewatering structures; at one point, a 'flame' of sandy clay passes up from the bed below and is truncated by the overlying clay. The pebbly horizon in the underlying sand varies from scattered small pebbles (up to 1 cm in diameter) to a 0.6 m bed with pebbles and cobbles up to 8 cm in diameter. Locally, the pebble bed is hard and cemented by laterite.

Deep excavations in the Henbury sand and gravel pit [SY 965 975] expose the top of the London Clay. At two places in the pit [SY 9631 9732; 9590 9725] bluish grey, finely sandy clay was dug from beneath 10 m of Poole Formation sands. In another part [SY 9613 9766], pink and grey clay, bedded in 1 cm-thick units, floors the pit and overlies dark grey, thinly bedded silty clay. Some 80 m east of this section [SY 9620 9765], interbedded thin beds of mottled pink and grey clay, and fine-grained sand occur in the lowest part of the pit. Dinoflagellate cysts from these clays [at SY 9631 9732; 9613 9776] include *Dracodinium simile* and *Homotryblium tenuispinosum*, the latter suggesting a lagoonal depositional environment.

North of the River Stour, the London Clay is divisible into mappable members. Springs commonly mark the base of the Warmwell Farm Sand, which, near Pamphill, forms a low scarp. Glauconitic sandy clay at the base of the sand was augered at one place [ST 9909 0102], east of the church.

North and east of Wimborne cemetery, clayey, orange-brown, fine-grained, silty sand up to 2 m thick forms a small feature. At High Hall around [SU 003 028], the Warmwell Farm Sand consists of 6 to 7 m of medium- to coarse-grained, orange-brown sand.

North-east of Higher Honeybrook Farm, pebbles were struck while augering in fine-grained sand at two places [SU 0143 0357; 0134 0362]. South-east of Stanbridge church, there is a broad outcrop of yellow and orange fine-grained sand, locally associated with mottled orange and grey silty clay [SU 0097 0363 and 0120 0386]. A pit [SU 0109 0387] 650 m east of the church shows the following section:

	Thickness m
Topsoil	0.50
WARMWELL FARM SAND	
Sand, fine-grained, moderately sorted, positively skewed, finely bedded, orange-brown, with small-scale (up to 2 cm high) cross-bedding,	0.90
Sand, fine-grained, silty, pale to medium grey	0.02 to 0.07
Sand, fine-grained, well-sorted, positively skewed, finely bedded, with some small (2 to 3 cm high) cross beds	
Sand, fine-grained, moderately sorted, positively skewed, pale grey	0 to 0.03
Sand, fine-grained, well-sorted, symmetrically skewed, finely bedded, with some small-scale cross bedding, buff; a few scattered clay clasts	0.50
Sand, fine-grained, well- to very well-sorted, positively skewed, cross-bedded in units 0.05to 0.2m high	0.90

The foresets in the basal bed are inclined at between 15° and 25° to the north-east. Coarse- to very coarse-grained sand was formerly worked in another old pit [SU 0140 0412], but no section remains.

Mottled orange and grey silty and finely sandy clay overlying the Warmwell Farm Sand was formerly worked for brickmaking near Holt [SU 0260 0385; 0293 0337]. There is a lenticular development of fine-grained sand within the clay near the school [SU 0200 0375] at Stanbridge. CRB

LYTCHETT MATRAVERS SAND AND OVERLYING CLAY

The clay above the Lytchett Matravers Sand was worked for the manufacture of bricks and pipes in a pit [SY 9485 9517] at Lytchett Matravers, where White (1917, p.18) noted 'a few feet of brown loam overlying a more sandy loam with nodules of ironstone'.

North of the village, fine-grained sandstone at the base of the member protrudes through the soil [SY 9445 9614]; sandstone at a similar stratigraphical level was worked for building stone in a pit [SY 9447 9609] 250 m to the east. A sample of the sand from west of the village [SY 9407 9523] is very fine-grained, well-sorted and symmetrically skewed. Ferruginous, fine-grained sandstone floors ditches [SY 9534 9572; 9543 9578] east of Lychett Matravers. CRB

Much laterite-cemented, fine-grained sandstone debris occurs in the fields [for example around SY 955 934] on the dip slopes north-west of Lytchett Minster. Large boulders of ferruginously cemented, fine-grained sandstone have been unearthed, and lie at the side of a field [SY 9512 9400] north-west of Post Green.

Mottled orange and grey silty clay caps the Lytchett Matravers Sand west of Race Farm [around SY 9527 9472] and near Post Green [SY 956 938], but at other outcrops around [SY 9585 9458; 9500 9307] the clay is grey and very stiff, with brick red mottles. ECF

The Lytchett Matravers Sand underlies the unconformable base of the Poole Formation on the north side of Wimborne [SU 013 009]. A sample from the top of the member [SU 0207 0125] consists of a very well-sorted, positively skewed, fine-grained, micaceous sand. It was in this area [c.SU 0138 0090] that White (1917, p.16) recorded a 'loamy sand with grains of glauconite'. North-east of Deans Grove, the fine-grained sand is very clayey [around SU 023 023 and 030 024]. Southwards, a mottled orange and grey silty clay, which thickens eastwards, is present within the sand. The upper leaf of the divided sand persists for about 1 km to the south-east before dying out.

The overlying clay crops out from beneath the Poole Formation in the area north of Merry Field Hill. It mostly consists of mottled orange and grey silty clay, but at two places [SU 0225 0177; 0237 0184] pinkish grey clay was augered.

Small (1 to 2 cm) rounded pebbles are common on the surface of the fields [SU 0325 0365] and in a ditch [SU 0332 0367] south-east of Holt church. A borehole [SU 0362 0407] east-north-east of the church proved the following sequence:

	Thickness m	*Depth* m
London Clay		
Clay, silty, yellow, becoming grey at 3 m	7.0	7.0
Clay, silty, and clayey silt, olive-grey	c.7.0	c.14.0
LYTCHETT MATRAVERS SAND		
Sand, very fine-grained, clayey, and silt with mica flakes	3.5	17.5
UNNAMED CLAY MEMBER		
Clay, silty, stiff, brownish olive-grey	3.0	20.5

Foraminifera from samples at 12.5, 15.5 and 20.5 m, identified by Mr M J Hughes, include:

	Depth		
	2.5 m	15.5 m	20.5 m
Alabamina westraliensis	common	—	rare
A. nobilis	very common	rare	common
Brizalina anglica	—	—	frequent
Cibicides lobatulus	frequent	rare	rare
Cibicidoides alleni	very common	frequent	common
Globigerina gp. *linaperta*	rare	—	—
Gyroidinoides octocameratus	rare	—	—
Karreria fallax	rare	—	—
Pullenia quinqueloba	rare	—	rare
Pulsiphonina prima	frequent	rare	rare
Spiroplectammina adamsi	—	rare	frequent
Turrilina brevispira	rare	—	—

The assemblages are consistent with King's (1981) Divisions B and C. Dinoflagellate cysts from samples at depths of 4.5, 10, 14 and 18.5 m include specimens of the *Apectodinium homomorphum* plexus, together with *Homotryblium tenuispinosum* at 10 m and *Dracodinium* sp. at 14 m; they indicate the *simile* Zone and a lagoonal environment of deposition. Thus, the combined foraminiferal and dinoflagellate cyst evidence indicates that these strata fall within either Division B or the basal part of Division C of King (1981, fig.45)

CRB

FIVE

Palaeogene: Bracklesham Group

INTRODUCTION

The Bracklesham Group comprises the Poole Formation and the overlying Branksome Sand Formation (Figures 8 and 9), and replaces the name Bournemouth Group, used in earlier reports (e.g. Bristow and Freshney, 1986b).

Although the Bournemouth Group of the district may not be the exact equivalent of the Bracklesham Group of the Southampton district (Edwards and Freshney, 1987b), there is sufficient correspondence to warrant a common name, of which Bracklesham has priority. The group in each area encompasses the strata between the top of the London Clay and the base of the Barton Group. In the Bournemouth district, the strata of the Bracklesham Group comprise dominantly fluviatile sediments, with only minor marine or estuarine deposits, whereas in the Southampton district, marine or estuarine deposits are dominant.

The base of the group appears to young south-westwards. In the Bournemouth–Poole area, the oldest dated deposits are of *coleothrypta* Zone age. In the Southampton district, the Wittering Formation, the oldest unit of the Bracklesham Group, is of *varielongituda* Zone age. However, the *coleothrypta* mircroflora may not be a strict chronostratigraphical indicator, but may in part be facies-controlled (Edwards and Freshney, 1987b, fig.7).

The Bracklesham Group is roughly equivalent to the Bagshot Beds plus the lower part of the Bracklesham Beds of the old One-Inch Geological Sheet 329 (Bournemouth). The Boscombe Sand, placed in the upper Bracklesham Beds by Gardner (1879; 1882) and in the Lower Bracklesham Series by Ord (1914), is here regarded as the basal formation of the overlying Barton Group. The Poole Formation equates essentially with the Pipeclay Series of previous authors (e.g. Ord, 1914; White, 1917), following Curry et al. (1978). The Branksome Sand includes the former Bournemouth Freshwater Beds and the Bournemouth Marine Beds.

Plint (1983b) referred to the strata between the London Clay and Barton Clay as the 'Bracklesham Formation', in which he recognised five sedimentary cycles. The evidence for the five cycles, each of which commences with a marine transgression represented by beach-barrier sands, is based largely on the more marine sequences exposed on the Isle of Wight, and it is now evident that these cycles cannot be directly correlated with the dominantly fluviatile sediments of the Poole Formation and Branksome Sand in the Bournemouth area. However, the four marine incursions represented by the lagoonal facies in the Poole Formation (Creekmoor, Oakdale, Broadstone and Parkstone clays) all predate Plint's transgression T^3 of the Isle of Wight (which correlates with the base of Prestwich's (1846) Bed 15). Plint placed this transgression imprecisely, because of a gap in exposure, below the Branksome Sand near Poole, and within the Poole Formation in the Wareham area (Plint, 1983b, fig.15).

ECF, CRB

POOLE FORMATION

Stratigraphy

The term Poole Formation was introduced by Curry et al. (1978, pp.21–22, table 1) for a sequence of sands and clays above the London Clay in the area south of Poole. The base is clearly defined at outcrop where the clean, medium- to coarse-grained sands of the Poole Formation rest on bioturbated, commonly micaceous clays of the London Clay or, in the subcrop, on the glauconitic, micaceous, fine-grained sands and colour-banded clays of the Christchurch Member of the London Clay. The top of the formation has been taken at the top of the Parkstone Clay Member.

The main part of the Poole Formation outcrop lies west of a line from Poole Harbour entrance, through Wimborne, to Holt Heath on the northern margin of the district (Figure 2). The Poole Formation consists of an alternating sequence of fine- to very coarse-grained, locally pebbly, cross-bedded sands, and pale grey to dark brown, carbonaceous and lignitic, commonly laminated clays. Locally, there are red-stained, structureless clays and silty clays. The sands are generally thicker than the clays and occupy just over half of the outcrop area. Five of the clay units, which have been extensively worked for the manufacture of bricks, tiles and pottery, have been named as members (Freshney et al., 1985). They are, in ascending sequence: the Creekmoor, Oakdale, Haymoor Bottom (only locally developed), Broadstone and Parkstone clays. This alternation of clay and sand units represents a series of marine transgressions during which lagoonal clays, and locally beach-barrier sands, were deposited, interrupting a mainly fluviatile sedimentary environment. The base of some of the major sand units is sharply defined on geophysical logs, and may be erosive.

The formation shows considerable variation in thickness. The most complete sequence, coinciding with an east–west downwarp, occurs near Poole, where the formation is about 160 m thick. South of Poole Harbour, the higher parts of the succession are mostly absent because of erosion, although the Broadstone Clay occurs at Arne. West of Poole, towards Lytchett Heath [SY 969 947], and north of Canford Heath, the sand below the Broadstone Clay oversteps the lower members of the Poole Formation to rest on the London Clay. The Poole Formation is only 30 m thick around Merley [SZ 015 983], and 40 m thick in the Knighton area.

In the eastern part of the district, the thickness of the formation at depth is known only from the Knapp Mill [SZ 1544 9380], Christchurch [SZ 2002 9301], Bransgore [SZ 1958 9505] and Hurn [SU 0999 0071] boreholes, where it is 78, 86, 79.5 and 67 m respectively. Geophysical logs of the Christchurch Borehole indicate that the Creekmoor Clay, and probably its underlying sand, are absent. Regional correlations (see Figure 12) suggest that the Creekmoor Clay is interdigitated with the upper part of the Christchurch

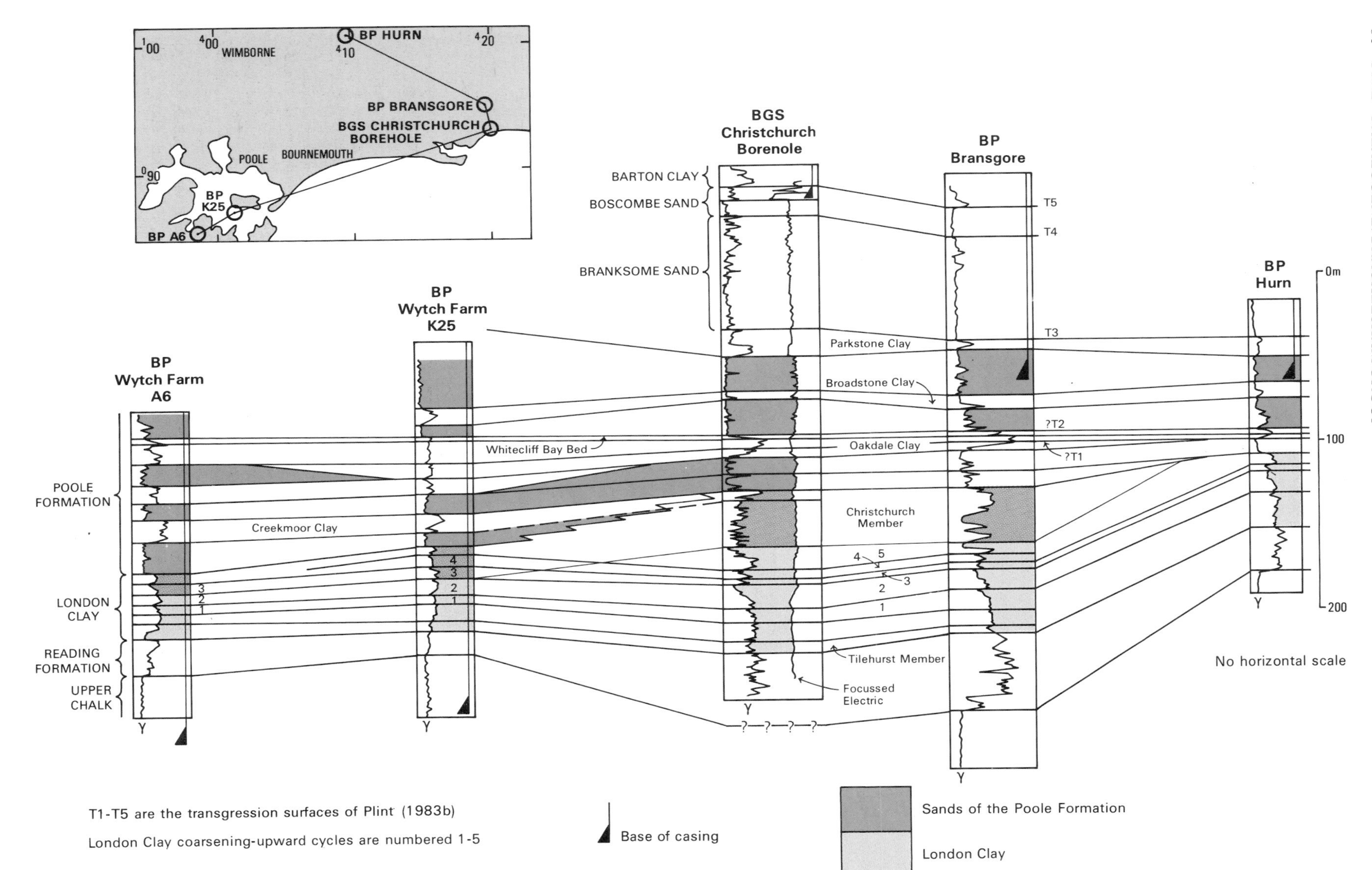

Figure 12 Gamma-ray log correlation of Palaeogene strata across the district.

Member of the London Clay beneath Poole Harbour. Near Hurn to the north, it is also probable that the Creekmoor Clay and underlying sand are missing, but there, the variation in thickness is the result of the northward overstep of the sand below the Broadstone Clay, which was probably caused by intra-Palaeogene tectonic movements and attendant erosion.

The grain-size distribution of the Poole Formation sands shows great variation (Figure 13C, D); the mean size ranges between coarse grained (0.2) and very fine grained (3.8ø), with an average close to the fine- to medium-grained boundary (2.0ø). The sorting ranges from very well-sorted (0.15ø) to poorly-sorted (1.8ø), with an average of 0.5ø. The symmetry of the distribution is also variable; the inclusive graphic skewness ranges from very positive (0.9) to negative (-0.3) with a positive average (0.2). The environmental interpretation of these properties is discussed on p.38.

Both fining- and coarsening-upward sequences occur in the Poole Formation. The former are both small scale (1 cm to 2 m) and large scale (10 to 20 m); coarsening-upward sequences are generally large scale. Both types of sequence are well displayed in geophysical logs, such as those of the Christchurch Borehole and the Wytch Farm oil wells; the interpretation of the logs is discussed on p.25. ECF, CRB

Creekmoor Clay and underlying sand

The Creekmoor Clay has a wide outcrop in the Creekmoor and Upton areas [SZ 000 935], where it was extensively worked for brickmaking; it is also thought to underlie much of Holes Bay [SZ 00 92]. The sand below it is known only from boreholes. Clay thickness recorded in boreholes ranges from 0 to 33 m; the thickness of the underlying sand ranges from 0 to 20 m. A representative section through the Creekmoor Clay and the underlying sand is given on p.40.

Where gamma-ray logs are available, as in the Wytch Farm Oilfield, the sand beneath the clay shows a reduction in clay content upwards. However, the top part of the sand shows a gradational contact with the overlying clay and fines upward into it.

At outcrop, the clay is usually off-white to pale grey, red stained and mottled, with subordinate dark grey, carbonaceous clay. In boreholes, dark brown carbonaceous clay, laminated pale grey and brown silty clay, structureless clay with listric surfaces, and clayey, fine-grained, laminated sand have been recorded. The fine-grained sand laminae in the laminated clays are usually discontinuous and lenticular. At one locality [SY 9809 9400] at Upton Heath there are signs of bioturbation.

Geophysical evidence, notably gamma-ray logs, and the limited dinoflagellate cyst evidence suggest that the Creekmoor Clay and its underlying sand are contemporaneous with approximately the upper 25 m of the Christchurch Member at Christchurch, the junction between the London Clay and Poole Formation being interdigitational beneath Poole Bay (Figure 12).

Oakdale Clay and underlying sand

The Oakdale Clay outcrop extends in an arc from the Arne Peninsula, through Holton Heath, Hamworthy, the southern part of Poole Harbour, Oakdale, Canford Heath and Creekmoor [SZ 005 942], to Upton Heath. The member varies rapidly in thickness from 0 to 45 m. The thickness of the underlying sand varies less rapidly and ranges from 5 to more than 30 m. Locally, the Oakdale Clay and underlying sand are missing.

The sand beneath the clay varies from fine to very coarse grained, and is locally pebbly; it is very well- to moderately-sorted (Figure 13C, D). In the Wytch Farm Oilfield boreholes, gamma-ray logs indicate both coarsening-upward and fining-upward sequences. In the Christchurch Borehole, on the other hand, the sand unit consists of several fining-upward sequences. The grain size and other characteristics of the fining-upward sequences (see p.38) suggest an alluvial depositional environment, whereas the deposits of the coarsening-upward sequences are probably mouth-bar or barrier-bar sediments (see p.39). Excellent sections of the sand beneath the Oakdale Clay can be seen in the old Marley Tile Pit [SZ 020 940] at Poole (see Freshney et al., 1985, pl.1). ECF, CRB.

At the type locality [SZ 021 932], the Oakdale Clay consists mainly of stiff, fissured, grey silty clay (see p.43). Elsewhere, the clay is commonly carbonaceous and laminated, and is red stained locally. In places, carbonaceous laminated clay passes laterally into red-stained grey clay.

To the north, north-east and south of Poole Harbour, the Oakdale Clay is divided into upper and lower leaves by up to 30 m of sand (Bristow and Freshney, 1986a; Freshney et al., 1985; Freshney and Bristow, 1987). The lower clay, the 'Arne Clay' of the ECC geologists working in the area south of Poole Harbour, crops out only in the Arne area; it is well exposed in the Arne clay pit [SY 975 896], where it consists of pale grey, structureless clay interbedded with brown lignitic layers. Elsewhere, red-stained layers have been noted.

Overlying the 'Arne Clay' is the 'green bed' which is used as a marker throughout the Wareham area by the ECC geologists. This bed consists of silty clays and sands, usually associated with a lignite, and locally with well-rounded black flint pebbles (personal communication, Q G Palmer). Mr Palmer also reports the local presence of glauconite grains both in the green bed and in an impersistent sand below it. The colour of the 'green bed' is partly due to ferrous iron; it is usually bright green on first exposure, but loses its colour within a matter of hours (Bristow and Freshney, 1987). The green bed is overlain by a dominantly arenaceous sequence, but clay and silty lignitic clay with, at least locally, a freshwater microflora occurs, particularly towards the base.

The upper leaf of the Oakdale Clay crops out at or near sea level in the Arne–Wytch Farm area. It comprises impersistent, grey, brown or black, carbonaceous, patchily red-stained, locally laminated clays. These clays, the 'Thrashers' or 'Newton clays' of the pit operators, occupy narrow linear channels [SY 970 839; SZ 010 846] within the underlying sands.

The Oakdale Clay in the Christchurch Borehole is represented by interbedded laminated clay and clayey, fine-grained sand, passing up into a structureless clay containing extensive rootlet networks at three levels. This upper clay correlates with the extensive Whitecliff Bay Bed, a major palaeosol in the Isle of Wight and Southampton areas (Edwards and Freshney, 1987a). In several boreholes, polished, striated shear planes, which are probably listric surfaces, characterise the palaeosols.

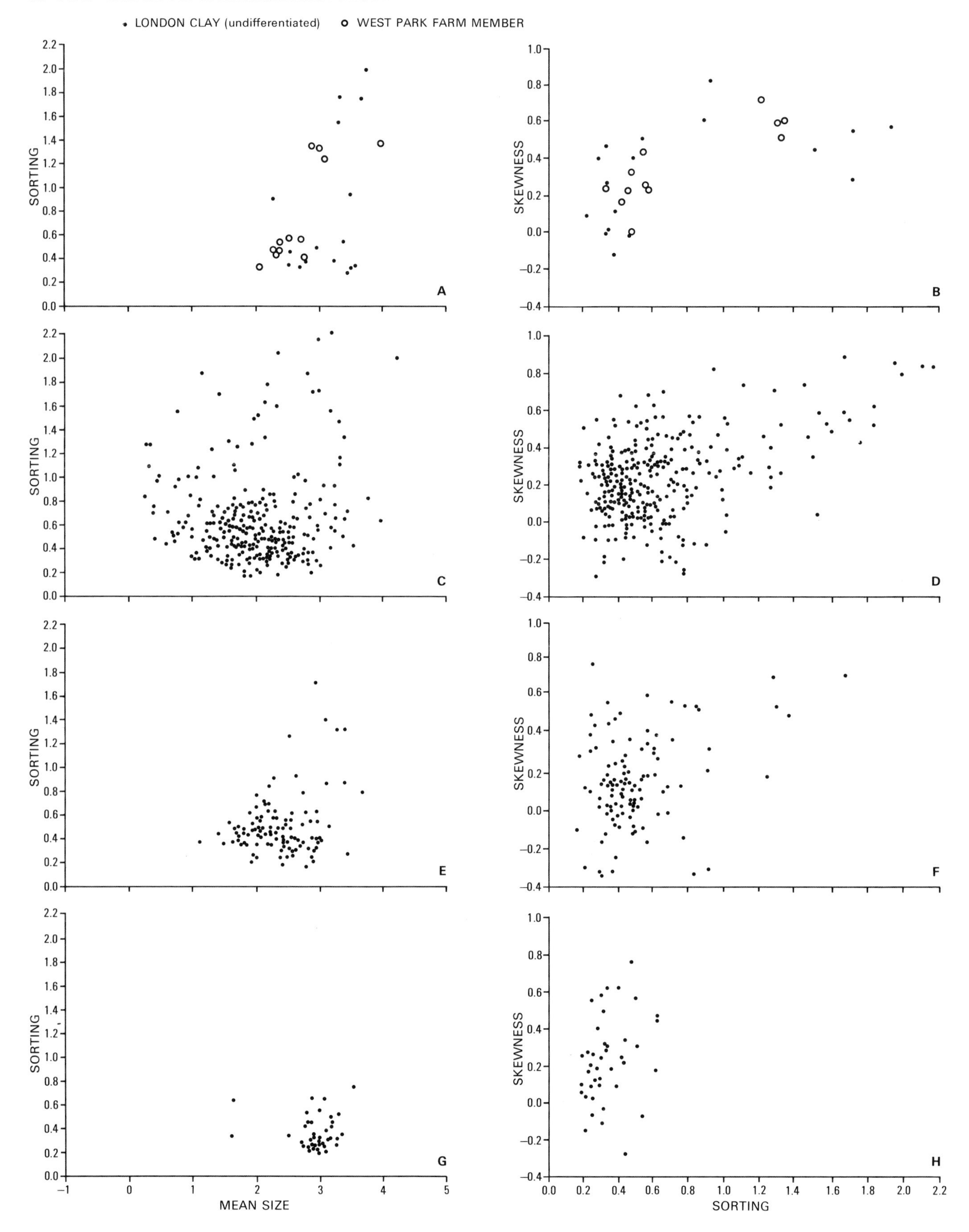

Figure 13 A–H. Scatter plots of grain size statistics of Tertiary sands of the Bournemouth district. (A–B, London Clay; C–D, Poole Formation; E–F, Boscombe Sand; G–H, Becton Sand)

Marine dinoflagellate cysts of *coleothrypta* Zone age have been recovered from several Oakdale Clay samples (Table 1) and indicate a probable tidal flat or saltmarsh depositional environment.

Haymoor Bottom Clay

The Haymoor Bottom Clay, consisting of up to 3 m of hard grey silty clay, crops out on low ground at Haymoor Bottom on Canford Heath [between SZ 026 947 and 030 940]. It rests on fine- to coarse-grained sand, which is lithologically indistinguishable from the sands above the clay. Much of the outcrop is hidden by drift and made ground. A lenticular clay up to 8 m thick, which may correlate with the Haymoor Bottom Clay, extends eastwards from Uddens Plantation [SU 059 019] into a northward-draining valley [SU 063 015]. ECF, CRB

Broadstone Clay and underlying sand

The Broadstone Clay takes its name from Broadstone [SZ 005 957], where the clay was worked for brickmaking [SZ 0085 9575] in the 1890's (White, 1917). This is the most widespread of the clays in the Poole Formation. In the north, the outcrop is locally impersistent, but it is continuous from Dudsbury to just south of Oakdale Cemetery [SZ 025 925]. West of Poole Harbour, the clay occurs on some of the higher ground at Holton Heath [around SY 945 910]. South of the harbour, it is present on the higher ground of the Arne Peninsula [SY 970 885]. On the Goathorn Peninsula, and near the top of the Wytch Farm No. F15 Borehole [SZ 0104 8574] sequence, the clay is missing, probably cut out beneath the sand which underlies the Parkstone Clay. The thickness of the Broadstone Clay varies rapidly from 0 to 19 m.

The sand beneath the clay is up to 30 m thick; it commonly has an erosional contact with the beds below and, in the northern part of the district, it rests directly on London Clay. South of Poole harbour, evidence from ECC Ball Clay Company's boreholes suggests that the sand cuts up to 30 m into the Oakdale Clay, and mapping indicates its presence in a 20 m-deep west-south-west-trending channel.

The sand is commonly cross-bedded, silty and dominantly medium grained, but ranges from very fine to very coarse grained (Figure 13C, D). On a small scale, there is much evidence of fining-upward units in the sands. In total, for example in the Christchurch Borehole, the sands show a coarsening-upward trend.

The overlying Broadstone Clay varies from pale grey silty clay, through homogeneous, medium grey silty clay, to laminated, lignitic, silty and fine-grained sandy clay. The grey clays are commonly extensively stained brick-red or lilac, especially in the Beacon Hill area. Marine dinoflagellate cysts indicative of the *coleothrypta* Zone have been found in several clay samples (Table 1).

Parkstone Clay and underlying sand

The type area of this member is Parkstone [SZ 036 910], Poole, where the clay was formerly extensively dug for the manufacture of pottery and tiles (Freshney et al., 1985). Most of the pits have now been filled, but poor exposures remain at some localities [e.g. SZ 0365 9097; SZ 0369 9133] (see p.50).

The sand beneath the Parkstone Clay locally rests discordantly on the Broadstone Clay, and cuts it out completely in the Goathorn area, south of Poole Harbour, and in the southern part of Poole. The sand, which may be up to 30 m thick, varies from silty and fine grained, to very coarse grained and pebbly; medium- to coarse-grained sand is the most common. Most of the sand is moderately sorted, but some medium-grained sand is well sorted, and some of the finer-grained sand is poorly sorted (Figure 13C, D). Several beds are graded, and fine upwards from a fine gravel at the base. In the Christchurch Borehole, the sand includes a 6 m coarsening-upward sequence between 135 and 129 m depth, above which it fines up to the base of the Parkstone Clay. The lower part of the clay shows a continuation of this upward fining. Impersistent breccias composed of clay clasts up to 10 cm across occur locally (Plate 3). The sand is well exposed in pits on Canford Heath [SZ 030 968] and Beacon Hill [SY 982 951], where common cross-bedding indicates the principal current directions to be from the south-west and west (Figure 14). The bedding is locally convolute (Plate 4). ECF, CRB

The Parkstone Clay crops out in the south on Brownsea Island and extends from Poole Head [SZ 050 884], through Parkstone, to the north of Alderney [SZ 040 950]. North and north-west of Alderney and eastwards from Canford Heath, the Parkstone Clay is cut out by the Branksome Sand. The clay reappears in the Stour valley and has a continuous outcrop as far north as Ferndown, where it is again cut out beneath the Branksome Sand; it re-emerges in the valley of Uddens Water.

The clay has been proved at shallow depth along Bourne Bottom [SZ 047 945 to SZ 063 935], and crops out in the parallel valley to the south-west [around SZ 058 933], where clay was formerly worked in large pits. The Parkstone Clay has been proved in a borehole [SZ 098 917] at Bournemouth between depths of 78.6 and 95.7 m, and in the Christchurch Borehole between depths of 95.49 and 110.60 m.

Thicknesses of between 14 and 22 m are common in the district, but on Brownsea Island, the Parkstone Clay is only 7 m thick. Thinner sequences probably result from downcutting by the Branksome Sand, which has an erosional base.

In general, in the type area, the Parkstone Clay comprises a lower medium grey to greyish brown and brown, plastic ball clay (as in the BGS Homark Borehole [SZ 0364 9096]), and an upper lignitic, commonly laminated clay, such as that seen in sections [SZ 0365 9127; 0390 9104] north and north-east of the Homark Borehole.

Biostratigraphy

A rich flora, indicating a tropical environment of deposition, was obtained by Chandler (1962; 1963) from the sand below the Broadstone Clay at Lake [SY 979 907] and Arne [SY 970 894], and from the Parkstone Clay at Poole Head [SZ 052 886]. No age can be deduced from it. Twenty samples of clay from the Creekmoor, Oakdale, Broadstone and Parkstone Clay members of the Poole Formation have yielded dinoflagellate cysts. All had sparse *coleothrypta* Zone assemblages indicative of a nearshore or lagoonal environment of deposition (Table 1). At Arne Clay Pit (Table 1, locality 50), the only known occurrence within the district of the nonmarine

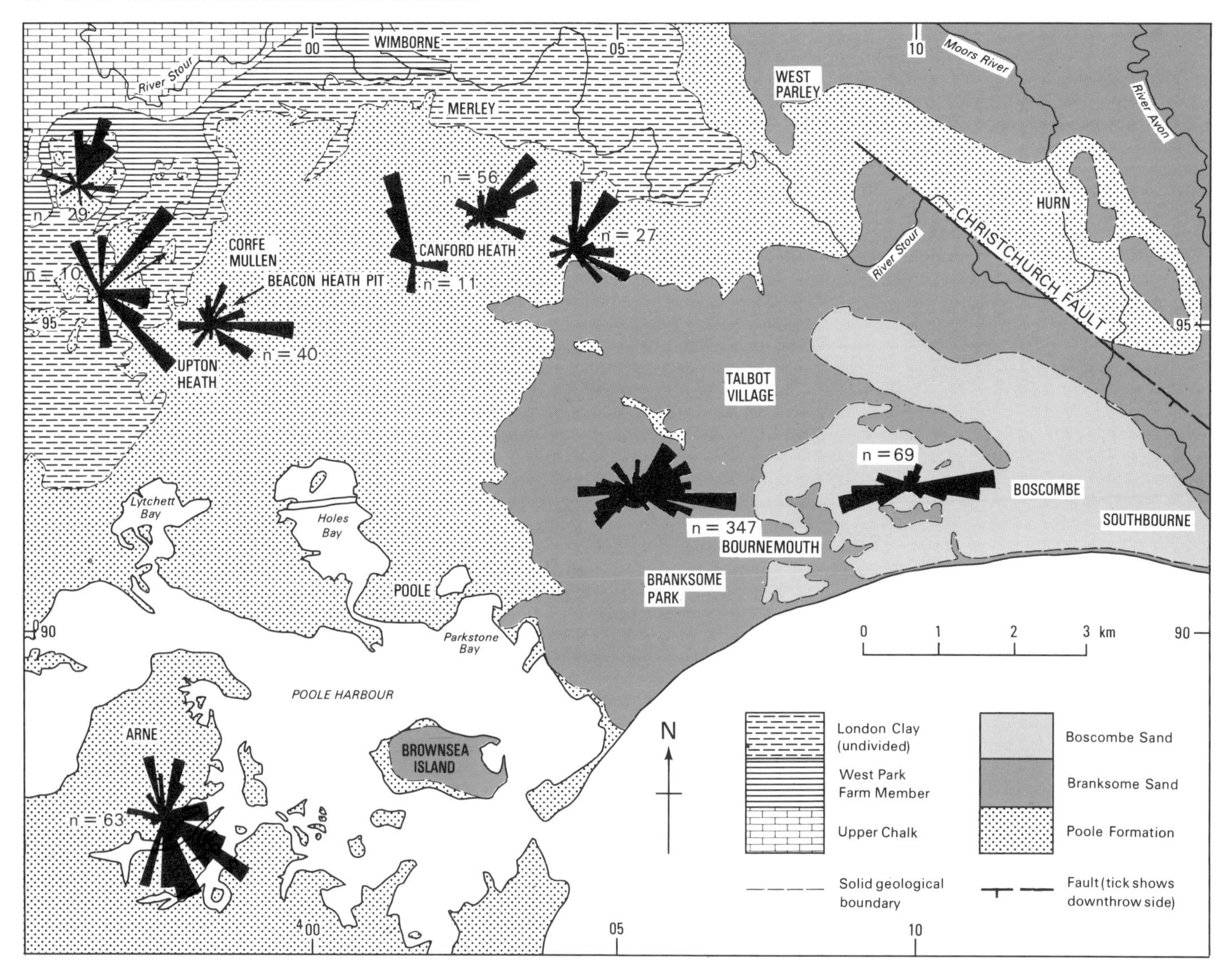

Figure 14 Current-rose diagram for sands of the Bracklesham Group/Boscombe Sand.

dinoflagellate cysts *Phthanoperidinium obscurum* is accompanied by freshwater algae known elsewhere only in the Headon Formation of the Southampton district (Edwards and Freshney, 1987b). ECF, CRB

Conditions of deposition

The basal beds of the Poole Formation reflect an abrupt incoming of fluviatile and coastal sands. The supply of sediment from the north-east, which characterised deposition of the London Clay, ceased and was replaced by a supply from the south and west (Morton, 1982). This was probably the result of contemporaneous faulting, which disrupted the regional sedimentation pattern.

Cross-bedding directions in the Poole Formation sands, though variable, indicate currents flowing dominantly from the west, but with some evidence of a flow from both north and south (Figure 14). The latter probably resulted from uplift in the area south of the Isle of Purbeck Monocline, which also probably led to the development of fault-scarp fan deposits at Bincombe Down [SY 688 853] and Blackdown [SY 613 875], farther west in the basin (Plint, 1982). The grading of the sands in fining-upward sequences, the presence of erosional and channelled surfaces at the base of many of the sands (Figure 15), and the ubiquitous occurrence of both trough and planar cross-bedding led Plint (1983a) to suggest that most of the beds in the Poole area in his Bracklesham Formation were of fluviatile or lacustrine origin. Similar sequences of fining-upward sands with clay-clast conglomerates at their bases occur in the Christchurch Borehole. However, some sand units in many of the Wytch Farm wells, and in the Christchurch Borehole, coarsen upward. One such unit in the Christchurch Borehole succeeds a major palaeosol and seatearth at the top of the Oakdale Clay. The basal sands of this coarsening-upward unit mark a marine transgression, and the overlying sands a probable prograding beach-barrier. There is another prograding sequence above this. Such sequences suggest the presence of

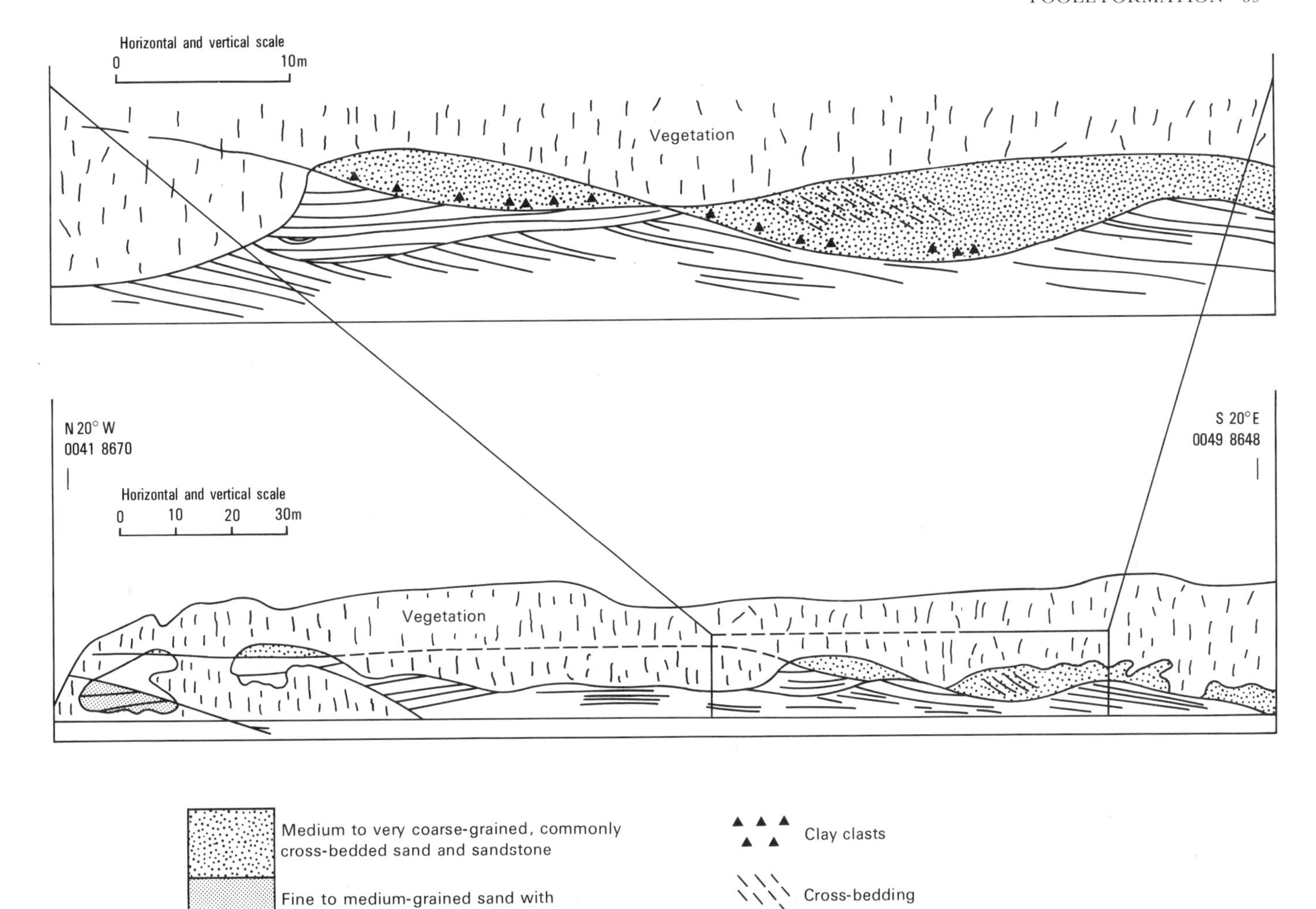

Figure 15 Relationship of coarse-grained sand with an erosional base to a sequence of rhythmic, muddy, fine-grained sands in the Poole Formation in cliffs on Green Island.

either mouth-bars or barrier-bars; this is supported by the occurrence of well-sorted and symmetrical to negatively skewed sands in the Poole Formation.

According to Chandler (1962), the flora, which has a humid tropical aspect, was largely that of a forest growing along river banks. Close by, dense colonies of sedges grew in pools, in quiet stretches of water and in marshes. Coastal marshes, only occasionally inundated at high tide, were colonised by *Acrostichum*, a fern tolerant of saline conditions.

The variable nature of the clay members of the formation suggests a range of depositional environments. Plint (1983a,b) considered them to represent the overbank deposits of a meandering river in a marsh or floodplain, the intervening sand members being the channel deposits. However, marine dinoflagellate cysts in all but the Haymoor Bottom Clay indicate that the environments were open to the sea, and that the deposits are probably of tidal-flat and salt-marsh origin. The clay members, which contain both probable fluviatile fining-upward sand sequences and more marginal, marine, coarsening-upward sand sequences, may have formed in tidally influenced back-barrier lagoons. They commonly show an overall upward grain-size reduction by the loss of sand. They contain palaeosols, some with red staining, indicating a fall in water level accompanying retreat of the lagoonal brackish waters, and sufficient drying out to allow a measure of oxidation in the sediment. Plint's 'pure clays' may have formed under such conditions. However, the clays mainly represent marine transgressions, which in some cases culminated in the formation of barrier sands in the Poole Harbour area.

Around Poole Harbour, the green bed within the Oakdale Clay probably represents the westernmost extension of a marine incursion, which can be tentatively correlated with the beds above the T2 transgression of Plint (1983b), and with the beds containing *Nummulites planulatus* (Bed IV) at Whitecliff Bay in the Isle of Wight.

Where best preserved, the Oakdale Clay continues up into medium- to fine-grained, well-sorted sands of possible beach-barrier origin, followed by back-barrier and swamp clays with a major palaeosol. This is abruptly followed by the fluviatile sands of the succeeding Broadstone sequence in the Poole area, and by barrier sands in the area around the Christchurch Borehole.

A complete, idealised transgression/regression cycle would reflect an upward change from fluviatile sands with some overbank clays and palaeosols, to laminated dinoflagellate cyst-bearing clays and silts, to beach-barrier sands, and back again through laminated clays to fluviatile sands. Transgression was never extensive enough in the Poole area to introduce fully marine conditions and the deposition of shelf sediments, except perhaps during the deposition of the 'green bed'. Such sediments do not generally occur any farther west than the central part of the Isle of Wight. Erosion during the fluviatile phases, coupled with probable tectonic activity, accounts for the incompleteness of some cycles.

The Poole Formation thus contains evidence of one major marine transgression, represented by the 'green bed' within the Oakdale Clay, and of minor marine incursions evidenced by the Creekmoor, Broadstone and Parkstone clays. Accompanying these events, belts of beach-barrier sands associated with back-barrier lagoons migrated backwards and forwards across a broad tract between the central Isle of Wight and the western Wareham Basin. ECF, CRB

Details

CREEKMOOR CLAY AND UNDERLYING SAND

Deposits near Upton House [around SY 995 928; 995 931] consist of brown carbonaceous clays and grey, red-stained clays. North-west of these localities, a roadside bank [SY 9928 9338] exposed 2.5 m of yellowish grey, silty, poorly laminated clay with abundant ferruginous layers. A borehole [SY 9994 9355] at Upton proved fine-grained sand above the clay:

	Thickness m
Poole Formation	
SAND BENEATH THE OAKDALE CLAY	
Sand, fine-grained, silty, yellowish brown, loose	0.76
CREEKMOOR CLAY	
Clay, firm, pale grey	1.22
Clay, with ferruginous-cemented siltstone at top	2.44
Clay, silty, pale grey, stiff	3.66

Another borehole [SY 9994 9303], east of Upton House, proved 1 m of silty fine-grained sand, with lenses of clayey silt and silty clay, below 5.1 m of lignitic clay.

Debris from flooded trial pits [SY 9889 9430] on the south side of Upton Heath comprised grey silty clay and blocks of ferruginous sandstone, probably derived from the overlying sand. An exposure [SY 9809 9400] in an old clay pit 80 m farther west shows:

	Thickness m
Poole Formation	
SAND BENEATH THE OAKDALE CLAY	
Sand, fine- to medium-grained	c.1.0
Clay, grey	c.1.0
Sand, medium- to very coarse-grained, slightly clayey	c.1.5
CREEKMOOR CLAY	
Clay, silty, grey to brownish grey and colour banded in shades of brown; very fine-grained sand and silt occur in pods subparallel to the bedding; occasional silt partings	1.5

ECF

Around Poole, the Creekmoor Clay and underlying sand vary in thickness from 30.5 to 36.9 m; within these beds, the clay/sand ratio shows marked variations. In boreholes, clay thicknesses vary from 9.9 [SZ 0137 9040] to 21.6 m [SZ 009 905], and the sand from 12.2 [SZ 0101 9181] to 22.3 m [SZ 0137 9040] (Freshney et al., 1985).

In the Dolphin Brewery well [SZ 009 905], the following beds in the Creekmoor Clay were recorded by the driller:

	Thickness m	*Depth* m
Poole Formation		
SAND BENEATH THE OAKDALE CLAY		
Sand, dark grey	15.24	59.13
CREEKMOOR CLAY		
Clay, sandy, pale grey	4.57	63.70
Clay, mottled red	3.05	66.75
Clay, black, loamy, peaty and sandy	1.52	68.27
Claystone	0.31	68.58
Clay, mottled	4.88	73.46
SAND BENEATH THE CREEKMOOR CLAY		
Sand, sharp, 'blowing', light coloured	2.44	75.90
Sand, sharp, 'blowing', dark	3.13	78.03
Sand, sharp, 'blowing', pale	0.91	78.94
Clay, black	0.92	79.86
Sand, sharp, 'blowing', pale grey	9.44	89.30
London Clay	12.20	101.50

The BGS Plessey Borehole [SZ 0058 9408] proved 4 m of yellow and red-stained grey clay (Creekmoor Clay) above 13.9 m of fine-grained, very clayey, lignitic sand. Shallow boreholes [SZ 0051 9386 to 0084 9384] along Cabot Lane proved more than 4 m of grey, brown and red clay and silty clay.

A borehole [SZ 0040 9347] in the floor of a 6 m-deep brickpit (now filled) proved 0.6 m of firm red clay above 3.6 m of bluish grey firm clay. Red-mottled, pale grey silty clay occurs in the banks of the partially filled brick pit [SZ 0008 9312] east of Creekmoor House, and in the fields [SZ 0003 9306] to the south. Boreholes farther east [SZ 010 933] near Fleet's Corner proved up to 12 m of mottled red and grey and brownish grey silty clay, stiff grey clay and very silty fine-grained sand [SZ 0074 9310], either at the surface or beneath tidal flat deposits to the south of Fleet's Corner. Boreholes at the sewage farm [around SZ 0085 9350] proved up to 9 m of stiff, grey, silty clay, locally brown and grey mottled and lignitic, beneath up to 2.6 m of sand.

Boreholes north of Fleet's Corner [around SZ 0113 9347] proved up to 2.7 m of sandy head, resting on up to 10 m of laminated sand, on white, grey, brown and red clays. A well [SZ 0110 9336] hereabouts proved, beneath 2.4 m of made ground: brown clay and sand, 3 m; grey clay with fine sand, 13.4 m; medium-grained sand with thin rock beds, 15.2 m; on London Clay.

North-east of Fleet's Corner around [SZ 012 938], many boreholes and several temporary sections in the Creekmoor Clay revealed (Freshney et al., 1985 p.30) dark brown, brownish grey and dark grey clays interlaminated with silt. In one borehole [SZ 0195 9438], the dinoflagellate cyst *Dracodinium similis* was recovered from a 2 m-thick clay bed. A BGS borehole [SZ 0127 9288] north of Stanley Green proved, beneath 17 m of sand, more than 3.5 m of sticky, red-stained, grey clay, which becomes lilac towards the base.

Reid (1894 MS) noted a section [SZ 0006 9245] on Pergins Island in 'gravelly wash above carbonaceous clay'.

Boreholes [SZ 0079 9142; 0063 9132; 0021 9096] in Holes Bay proved up to 7.7 m of estuarine deposits, resting on to 16.7 m of laminated, fissured, slickensided (i.e. with listric surfaces), stiff to very stiff, pale grey, greyish brown, dark brown, or mottled pinkish red and grey, lignitic, silty clay, locally grading to clayey silt; resting on 0.3 m of slightly gravelly, fine- to medium-grained, slightly silty sand.

Creekmoor Clay, variously described as blue, grey, brown or red-mottled, underlies Lower Hamworthy, and is concealed by between 3.6 and 4 m of river terrace deposits, and between 0.3 and 4.7 m of the overlying sand. The clay surface dips eastwards, and the overlying deposits increase to at least 15 m [e.g. in a borehole at SZ 0082 9008].

On the Arne Peninsula, the Creekmoor Clay has been proved in boreholes. Its thickness ranges from 7 to 33 m; the underlying sand varies from 10 to 25 m in thickness. ECF, CRB

OAKDALE CLAY AND UNDERLYING SAND

Sand outliers near Lytchett Matravers

Near Lytchett Matravers and Combe Almer [SY 945 978], the sand beneath the Oakdale Clay consists of up to 10 m of dominantly medium- to very coarse-grained sand. Northwards, it cannot be distinguished from the sand beneath the Broadstone Clay, and it is probable that the basal sand of the Poole Formation in that area is an amalgam of the two sands; it rests unconformably on London Clay. The base of the sand is commonly marked by springs.

Poor exposures of up to 1.5 m of orange, medium- to coarse-grained sand occur in two pits [SY 9504 9518; 9535 9520] east of Lytchett Matravers. Angular blocks of very fine-grained grey silcrete occur locally on the surface [SY 9435 9415].

Outcrops north of Poole Harbour

The 5 m-high scarp which extends south-eastwards from Pike's Farm [SY 9405 9303] consists mostly of coarse-grained sand, but locally, fine-grained and very coarse-grained sand occur. South of Pike's Farm, there is an extensive outcrop, up to 1.5 km wide, of mottled red and grey, pale or medium grey and mottled yellow and grey, locally carbonaceous clays and silty clays. CRB

In a pit [SY 9515 9215] north-west of Holton Heath, the Oakdale Clay is over 11 m thick, and consists mainly of plastic red and lilac-stained clays with some 2 m of grey and blue clay. Clay from this pit was worked for ball clay (Highley, 1975) until 1978.

In the north-western part of Holton Heath, the Oakdale Clay, about 3 to 4 m thick, consists of yellowish grey sticky clay and silt, with lateritic debris. Reid (MS, BGS) noted carbonaceous brown clay at Holton Clump [SY 9672 9135].

In the ECC Ball Clay Co's Beacon Hill Borehole [SY 9761 9446], the Oakdale Clay appears to be represented by only 1.5 m of sandy clay, some of which is red stained and some brown and lignitic.

In the Upton–Creekmoor area, the Oakdale Clay contains a sand interbed. The upper clay is slightly carbonaceous and laminated, while the lower is grey with red and lilac stains. The following section [SY 9972 9488] occurs west of the old railway:

	Thickness m
Poole Formation	
SAND BENEATH THE BROADSTONE CLAY	
Sand, very clayey, pebbly, with ferruginous fragments	0.6
OAKDALE CLAY	
Clay, brownish grey, roughly fissile	1.5
Laterite, brown and orange, impersistent	0.03
Sand, medium- to coarse-grained, yellowish orange, with a ferruginous layer (2–3 cm) at top; becomes very coarse-grained downwards, with many flint grains; rather clayey at base, with pebbles and clay clasts	4.0
Clay, grey, silty, containing a 2 cm bed of very fine-grained sand	0.3
Sand, fine- to very fine-grained, with layers of clay/silt	1.2
Sand, fine- to medium-grained, yellow	0.5
(The fine-grained sand and clay units above pass laterally south-eastwards into the coarse-grained sand)	
Clay, pale to medium grey with red and lilac mottles	3.5
Sand, fine- to coarse-grained, yellowish orange	1.0

An old pit [SY 9947 9402] at Creekmoor exposes (Plate 1):

	Thickness m
Poole Formation	
OAKDALE CLAY	
Clay, mottled grey and orange, silty; hard ferruginous layer at base.	0.8
Clay, silty, brownish grey, interlaminated with very fine-grained sand; the clay layers range from 2 to 5 mm thick; the sands range between 1 and 5 mm thick, but pod-like lenses occur up to 80 mm thick; some beds of clay and sand show intense disturbance, in zones about 20 to 30 mm thick.	0.5
Clay, silty and sandy, shows a poorly defined banding in siltiness; passes into a clayey very fine-grained sand at base.	c.1.0
Sand, very clayey, fine-grained, with variable amounts of clay in beds and laminae ranging from 0.5 to 9 mm in thickness; clay seams have sharp bases; some cross-cutting clay and sand units	0 to 1.0
Clay, in fairly regular units 8 mm thick, with thin, impersistent, fine-grained sand layers up to 2 mm thick, which have sharp bases and which fine upwards	0 to 1.5
SAND BENEATH THE OAKDALE CLAY	
Sand, medium- to coarse-grained, cross-bedded.	2.0

Many sections in the Oakdale Clay and the sand beneath occur in a disused clay pit on Upton Heath. The following [SY 9877 9428] is typical:

	Thickness m
Poole Formation	
OAKDALE CLAY	
Clay, greyish yellow	c.1.0
Laterite, orange-brown	0.2 to 0.3
Sand, fine-grained, rather clayey	c.2.0
Sand, buff, clayey	c.1.5
Ferruginous layer, rather soft	0.1
Clay, brownish grey to greyish brown, with plant debris	1.0
Silt, clayey, greyish brown to brownish grey, roughly laminated and bedded	2.0
Sand, fine-grained, buff, clayey, ferruginous layer at base	1.5
Clay, very silty and greyish brown, with fine plant debris	3.0

Plate 1 Creekmoor Sand Pit. Fining-upward microcycles in the Oakdale Clay. The base of each cycle, a very coarse-grained sand, rests with a sharply defined base on the clay of the underlying cycle.

The lower part of the clay, passing into the sand below, is exposed in a nearby section [SY 9875 9421]:

	Thickness m
Poole Formation	
OAKDALE CLAY	
Laterite, hard, brown	0.02
Silt, greyish orange, clayey, roughly laminated	2.00
Silt, very clayey, and extremely silty clay, roughly laminated; clay and silt well-laminated at base	2.20
SAND BENEATH THE OAKDALE CLAY	
Sand, medium- to coarse-grained; ferruginously cemented in top 2 m; flint grains common	2.00

Exposures [SY 9900 9330 and 9825 9383] north-west of Upton House reveal up to 8 m of fine- to medium-grained, locally cross-bedded sand. In the south-western part of Upton, a borehole [SY 9754 9306] penetrated the following sequence beneath 0.4 m of made ground:

	Thickness m
Poole Formation	
SAND BENEATH THE OAKDALE CLAY	
Sand, medium-grained, with some cementation	1.2
Sand, medium-grained, silty, dark brown, medium dense	0.2
Sand, coarse- to medium-grained, pale yellowish brown, dense, with some cementation	1.1
Sand, coarse- to medium-grained, clayey, pale yellowish brown, with some cementation	0.6

A borehole [SY 9827 9184] at Hamworthy proved the following:

	Thickness m	*Depth* m
HEAD		
Sand, brown and white	1.12	1.12
Poole Formation		
OAKDALE CLAY		
Sand, brown, with a little clay	0.10	1.22
Ball clay, grey	1.12	2.34

Clay, stiff, greyish brown	0.81	3.15
Sand and clay, brown	0.07	3.22
Clay, stiff, greyish brown	0.79	4.01
SAND BENEATH THE OAKDALE CLAY		
Sand, limonitic	0.38	4.39
Sand, hard, limonitic, with some grey clay	0.21	4.60
Sand, reddish brown, with some clay	1.24	5.84
Sand, lignitic, with clay	0.10	5.94
Sand, grey	0.26	6.20

A section [SY 9739 9108] near Rockley Sands shows:

	Thickness m
Poole Formation	
SAND BENEATH THE BROADSTONE CLAY	
Sand, coarse- to very coarse-grained, with some thin pipe-clay layers; cross-bedded from WSW and SW;	1.5
Sand, medium- to coarse-grained, very coarse-grained at the base where it is also ferruginously cemented	c.3.0
OAKDALE CLAY	
Clay, grey, silty, apparently structureless	0.5
Clay, brown, carbonaceous, with some lignite.	1.0

Dark brown carbonaceous clay containing patches of pale grey, red-mottled clay occurs below beach shingle at Rockley Sands [SY 9738 9093]. The dinoflagellate cysts *Kisselovia* cf. *coleothrypta* and members of the *Apectodinium homomorphum* plexus were recovered from the clay.

North-east of Darby's Corner, a borehole [SZ 0141 9482] passed through 5 m of sand and made ground into 9.5 m of medium grey, stiff clay, with patchy red staining; this clay becomes brown and highly lignitic with depth and rests on 3.1 m of laminated, pale grey and dark brown clay, which in turn overlies more than 3 m of coarse- to very coarse-grained sand.

Near the Grammar School [SZ 0141 9482], the Oakdale Clay is 12.6 m thick and rests on more than 30 m of coarse-grained sand. Clay, variously described as stiff, mottled red, brown, grey and white, silty clay and silt, together with firm, grey, black-mottled sandy clay, was proved in boreholes in the Darby's Corner area (Freshney et al., 1985, p.35). A section [SZ 0246 9463] at the roundabout east of Darby's Corner showed:

	Thickness m
Poole Formation	
OAKDALE CLAY	
Clay, greyish brown to brown, with pale grey silt wisps, some brecciated areas and, locally, siderite concretions	1.5
Sand, pale grey, medium-grained, very well-sorted and symmetrically skewed, in units of 2 to 5 cm; some cross-lamination showing current direction from the west; thin impersistent carbonaceous clay seams (1 – 5 cm); tree trunk reported in recumbent position in the sand; many carbonaceous layers in the lowest 0.8 m, where cross-bedding dips 20° south-eastwards	2.0
Clay, dark brown, carbonaceous, laminated, with small lenses of fine-grained sand 2 to 3 mm thick by 20 mm long	1.2

A borehole [SZ 0245 9445] near this locality proved:

	Thickness m	*Depth* m
Topsoil	0.6	0.6
Poole Formation		
OAKDALE CLAY		
Clay, firm, brown, mottled grey	1.2	1.8
Clay, stiff, brown, mottled grey	0.8	2.6
Clay, stiff, grey, interlaminated with fine-grained sand; small 'shell' fragments at base	2.6	5.2
SAND BENEATH THE OAKDALE CLAY		
Sand, fine- to coarse-grained, pebbly, with small lignite fragments	1.8	7.0

Clay samples from the roundabout [SZ 0246 9463] and a nearby borehole [SZ 0248 9430] yielded *coleothrypta* Zone dinoflagellate cysts.

Excellent sections in up to about 10 m of the sand beneath the Oakdale Clay occur in the Marley Tile Pit [SZ 020 940], east of Waterloo. About 4 m of coarse- to very coarse-grained or pebbly lignitic sand is exposed in the northern part of the pit [SZ 0198 9420]. The sand is cross-bedded (planar to concave-up), in sets and cosets between 0.1 and 0.8 m thick. The lignite occurs as disseminated fragments and as rounded pebbles up to 5 cm across. About 75 m to the south-east [SZ 0202 9415], 1 m of medium- to coarse-grained, and some very coarse-grained, orange sand with dewatering structures, overlies 5 m of greyish brown, very coarse-grained, well-bedded, cross-bedded lignitic sand. Higher beds in the south-eastern part of the pit comprise finer-grained, yellow, very well-sorted sands. These sands are overlain by 0.4 m of brownish grey clay of the Oakdale Clay. The basal part of the Oakdale Clay, at a depth of 1.9 m in a borehole [SZ 0195 9438] hereabouts, yielded the dinoflagellate cyst *Dracodinium similis*.

A borehole at Oakdale [SZ 0217 9323] penetrated the full thickness of the clay and part of the underlying sand:

	Thickness m	*Depth* m
Poole Formation		
SAND BENEATH THE BROADSTONE CLAY		
Sand	8.6	8.6
OAKDALE CLAY		
Clay, silty, grey, very stiff, fissured	5.0	13.6
Lignite	1.5	15.1
Clay, silty, grey, very stiff to hard, fissured	7.9	23.0
SAND BENEATH THE OAKDALE CLAY		
Sand, grey, very dense	2.0	25.0

A nearby borehole [SZ 0213 9322] proved 17.5 m of clay; the uppermost 3.2 m were described as white silty clay. In boreholes south of the old brick pits at Oakdale, the clay is described in drillers' logs as 'dark brown laminated clay' [SZ 0205 9380] and 'firm dark blue clay' [SZ 0217 9379], beneath 1 to 2.1 m of mottled yellow, brown and grey clay.

At the Transfer Station on the Nuffield Estate [SZ 0195 9375], up to 4.7 m of stiff to hard, brown and dark grey, laminated, silty clay has been proved in boreholes to rest on lignitic, moderately sorted, medium- to coarse-grained sand.

Boreholes around [SZ 017 924], an auger hole [SZ 0155 9214] and stream bank exposures [SZ 0149 9213; 0123 9215] east of Stanley Green revealed respectively, grey and dark brown silty lignitic clay, pale grey silty clay, and yellow and red mottled clay.

Boreholes along Willis Way around [SZ 009 923] proved silty clay, firm mottled reddish brown and grey in its upper part, and bluish grey and laminated in its lower part, above fine- to coarse-grained, silty, micaceous and lignitic sand.

The BGS Sterte Borehole [SZ 0095 9200] proved 8.5 m of brown carbonaceous Oakdale Clay, interlaminated with fine-grained sand, beneath 6.5 m of made ground and estuarine deposits. The underlying deposit, 20.4 m thick, varies from a clayey silt to a very coarse-grained lignitic sand. Dinoflagellate cysts indicative of the *coleothrypta* Zone were obtained from the clay at a depth of 11.25 m. About 300 m south-west, a trial hole [SZ 0076 9174] encountered more than 13.5 m of laminated silty clay and clayey silt, with thin siltstones. The uppermost 3 m were mottled and streaked with pink and red; the lower beds were dominantly pale grey, but some were dark brown. Listric surfaces were noted in some cores.

At Poole hospital [SZ 0180 9140], the thickness of the clay and underlying sand are 10.4 and 16.1 m respectively. Beneath Poole, the Oakdale Clay comprises upper and lower units separated by a lenticular bed of sand, which is up to 30 m thick; the upper clay is at least 19 m thick, and the lower 10 to 17 m.

Boreholes [around SZ 0145 9130; 0140 9115] north of Poole Station, which commenced in the upper clay, proved between 8 and 17 m of stiff silty clay and clayey and sandy silt, varying from pale grey to dark brown, with some purplish staining; resting on fine- to coarse-grained lignitic sand. Black pebbles, which occur in a silty sand between depths of 20.1 and 21.6 m in one borehole [SZ 0149 9116], probably represent the T2 Transgression of Plint (1983b). A borehole [SZ 0137 9040] at the old gasworks proved 30 m of sand above 9.9 m of the lower clay unit. Pebbles were noted in the top of the clay, which was overlain by 1.8 m of 'green' sand (Whitaker and Edwards, 1926). The green sand and pebbles again probably represent Plint's T2 Transgression.

Quaternary deposits overlie the upper clay under much of the old part of Poole. One of the thickest clay sequences is in a borehole [SZ 0120 9067] in the High Street, where 3 m of made ground and river terrace deposits overlie more than 9 m of stiff, dominantly pale grey, laminated, silty clay.

Several boreholes in Lagland Street [around SZ 0145 9045] proved the Oakdale Clay and the underlying sand. In one [SZ 0147 9043], the clay consisted of more than 4.5 m of greyish brown silty clay; the sand, more than 11 m thick, is described in drillers' logs as fine to medium grained. East of the gasworks, boreholes [e.g. SZ 018 905] proved up to 3.6 m of firm, brown, laminated clay and silt. Boreholes [around SZ 009 909] south-west of the Lifeboat Station, Poole, proved the lower clay beneath 1.9 m of made ground, comprising 16.1 m of very stiff, pale grey, red-speckled silty clay, with grey, orange and brown clayey silt between 15.8 and 17.3 m depth; the basal 6.7 m were laminated.

The lower leaf of the Oakdale Clay was proved in boreholes beneath made ground, estuarine deposits and the overlying sand on the northern margin of Poole Harbour at depths of between 10 [SZ 0064 8993] and 23 m [SZ 0087 8970]. The clay, up to 16 m thick, is commonly described by drillers as firm to stiff, grey and pale brown, laminated, lignitic and silty, with many 'polished, striated, shear planes' (listric surfaces); patchy reddish brown mottling was also noted. ECF, CRB

Outcrops south of Poole Harbour

The upper leaf of the Oakdale Clay crops out extensively on the foreshore of the Arne Peninsula, and along several of the valleys which drain it. However, head deposits associated with springs issuing from the overlying sand, obscure many of the outcrops. In one area, on Slepe Heath [SY 9440 8615], the overlying sand is cambered over the clay.

A section [SY 9575 8709] on the north-east of Slepe Heath showed:

	Thickness m
Poole Formation	
SAND BENEATH THE BROADSTONE CLAY	
Sand, coarse- to very coarse-grained, with a ferruginously cemented base	3.0
OAKDALE CLAY	
Clay, brown, silty, with thin silt laminae	1.1
Clay, pale grey, with laterite beds 2 to 3 mm thick	1.1
Sand, medium-grained, yellow, with a lateritic top	0.2

The basal sand recorded above is probably a sand bed within the Oakdale Clay.

Pale to medium grey silty clay is exposed in a low cliff on the south-west side of Arne Heath [SY 9566 8770]. From that point, the clay dips down below beach level to reappear as a brown silty clay about 200 m to the north-east. In the cliff [SY 9606 8845] west-north-west of Arne, there is a lens, 2 m thick, of brownish grey, laminated, silty clay within coarse-grained yellow sand. The clay yielded one specimen of the dinoflagellate cyst ?*Kisselovia* sp. indet. indicative of the *coleothrypta* Zone or younger. Greyish brown silty clay occurs at the foot of the cliff [SY 9638 8889] to the north. Up to 2.5 m of medium grey, shaly, locally strongly cryoturbated clay was seen in the cliff [SY 967 891] about 350 m north-east of the latter occurrence.

The lower part of the Oakdale Clay, much obscured by drift deposits, appears towards the tip of the Arne peninsula, where it is worked in a large pit [SY 976 896] for ball clay (Plate 2). A section [SY 9758 8961] in the middle of the pit, where the overburden had been stripped off, showed:

	Thickness m
Poole Formation	
OAKDALE CLAY	
Clay, pale grey, structureless, slickensided, with orange-stained joint surfaces	3.0
Clay, dark brown, with lignite fragments; diffuse contact with beds above and below	0.6
Clay, pale grey, structureless	2.2
Clay, brown, with three equally spaced dark brown layers; diffuse contacts top and bottom	0.6
Clay, pale grey, structureless	2.2

The clays rapidly weather orange on exposure to air.

A section [SY 9753 8953] on the south side of the pit showed 0.6 to 1.0 m of peat, above 2 m of buff-orange, fine-grained sand, in turn overlying 1 m of greyish brown, laminated, lignitic clay. Another section [SY 9756 8951] close by showed the topmost 0.5 m of the clay to be finely laminated, dark grey and silty, with a thin bed of coarse-grained sand at its base, together comprising the 'green bed' of the pit operators. The basal sand thickens rapidly eastwards to 1.8 m, and is still overlain by the greyish brown lignitic clay. A sample from this clay contains many examples of the freshwater alga *Pediastrum* sp. and the nonmarine dinoflagellate cyst *Phthanoperidinium obscurum*.

Plint (1980) recorded the following section in this pit:

	Thickness m
Poole Formation	
OAKDALE CLAY	
Clay, silty, orange-red, with discontinuous sideritic nodules	1.50

Plate 2 Ball clay in the Oakdale Clay, Arne. An alternating sequence of pale grey, structureless, slickensided clay (ball clay) and carbonaceous lignitic clay.

Silt, sandy, pale grey, bioturbated to structureless, with trains of ripples	1.10
Sand, fine-grained, with straight-crested ripples on top; internal climbing ripples and ?stem casts; strongly loaded base	0.70 to 0.90
Clay, silty, thinly interbedded with grey-green silty sand; thin bands of sideritic nodules; weak bioturbation	1.80
Clay, silty, laminated, grey-green, microconvoluted and microfaulted throughout ('green bed') 0.50	
Clay, silty, dark grey, laminated and burrowed; white silt burrowfills with meniscus structure	0.75
Clay, pale grey, silty and sandy at base, resting with a sharp, slightly bioturbated, interface on	1.35
Clay, white, pure [ball clay]	0.55

Mottled orange and grey clay of the upper leaf of the Oakdale Clay occurs at shallow depth on the flanks of Arne Bay [SY 977 893 and 979 894]. On the north-eastern side of Long Island, grey silty carbonaceous clay at beach level [SY 9872 8811] may be the Oakdale Clay (see p.48). CRB, ECF

Some 1.5 m of grey and brown carbonaceous clay crops out on the foreshore [SZ 0215 8553] of Brand's Bay. From that point north-eastwards, grey clay is intermittently exposed along the shore to just south-east of Jerry's Point [SZ 0302 8587], beyond which the outcrop is covered by Quaternary deposits. Typical exposures [SZ 0281 8590; 0280 8614] show stiff, grey clay below high-water mark, and red-stained grey clay above. Another section [SZ 0290 8606] shows carbonaceous clay below high-water mark. Inland, yellow and grey clay, locally reddened, occurs beneath gravelly and clayey sand. ECF

HAYMOOR BOTTOM CLAY

The Haymoor Bottom Clay, up to 3 m thick, is variously described in drillers' logs of boreholes [SZ 0264 9466; 0308 9461; 0321 9432] and in sections [SZ 0295 9460] on Canford Heath, as hard or stiff, grey or greyish brown, very silty and sandy clay, or clayey silt (Freshney et al., 1985, pp.41–42). CRB, ECF

A possible correlative of the Haymoor Bottom Clay, up to 8 m thick, extends eastwards from Uddens Plantation [SU 059 019]. The clay is fairly stiff, grey, locally mottled reddish brown, and usually finely laminated, and contains thin pale silty and sandy interbeds. At least 9.5 m of fine- to medium-grained sand, with thin clayey beds, occurs below the clay [SU 0665 0183] south-west of Ameysford. BJW

Broadstone Clay and underlying sand

Outcrops north of the River Stour

North of the River Stour, the sand underlying the Broadstone Clay varies from fine- to coarse-grained, but is dominantly fine grained, commonly well sorted, negatively to symmetrically skewed and locally ferruginously cemented. Sections in the upper part of the sand commonly show thin (0.05 to 0.15 m) clay interbeds. The base is almost everywhere marked by springs. The Broadstone Clay consists mostly of mottled orange and grey silty clay.

Near White Sheet Plantation in the north, a ditch section [SU 0512 0291] revealed 1.5 m of yellow silty clay overlying 0.4 m of grey and yellow sandy clay. North of Uddens Water, as well as mottled orange and grey silty clay, pale grey silty clay [SU 0433 0267], mottled red and yellow clay [SU 0452 0279], pinkish brown silty clay [SU 0451 0320] and dark grey, fine-grained sandy clay [SU 0444 0454] also occur. BJW, CRB

Just east of Pilford, much ferruginously cemented fine- and medium-grained sand occurs on the surface. One sample of sand [SU 0352 0159] is fine grained, well sorted and negatively skewed. Beds of coarse- to very coarse-grained sand occur, for example, north-west of Bedborough Farm [SU 0445 0215].

Several sections in the underlying sand were exposed during the construction of the Ferndown Bypass. One section [SU 0492 0093] in the top part of the sand showed:

	Thickness m
Poole Formation	
Sand beneath the Broadstone Clay	
Clay, sandy, mottled orange and grey	0.4
Sand, fine-grained, orange	0.1
Clay, sandy, mottled orange and grey	0.1
Sand, fine-grained, orange	0.2
Clay, sandy, mottled orange and grey	0.05 to 0.15
Sand, fine-grained, orange-brown	0.3

About 130 m south-west, a section [SU 0483 0086] revealed the junction with the Broadstone Clay. On the south side of the cutting, there is up to 1.5 m of orange-brown clay with a gently undulating base which has an overall dip of about 2°NE. On the north side, the clay dips steeply northwards and passes into a dark grey lignitic and ?marcasitic clay. The clay overlies mottled orange and grey, very clayey, fine-grained sand, which passes north-eastwards into a fine-grained orange sand.

In a third section [SU 0472 0076] 140 m to the south-west, the Broadstone Clay/sand contact dips 15°NE. There, more than 3 m of pale grey clay, silty clay and fine-grained sandy clay, with thin layers of laterite in the basal 2 m, overlie more than 3 m of orange-buff, well-sorted, negatively skewed, fine-grained sand. Irregular pods of sand up to 2 m thick occur in the clay. CRB

A borehole [SU 0533 0131] at Stapehill penetrated 4 m of mottled grey and brown silty clay, above 6 m of dark grey silty clay, thinly interlaminated with pale brown silt and sand. BJW

A section [SU 0482 0052] in an old railway cutting showed 3 m of orange-brown, finely bedded (in units 1 to 5 mm thick), medium-grained sand with some small-scale (up to 8 cm high) cross-bedding; lenses of grey silt up to 10 mm thick occur locally.

The Broadstone Clay on the south side of Colehill varies in thickness from about 8 to 14 m, with little lithological variation. On the outskirts of Wimborne, 1.3 m of mottled pink and brownish grey silty clay were seen in a temporary exposure [SU 0167 0055]. Pinkish brown silty clay was also augered in the upper part of the old Wimborne brick pit [SU 0187 0100]. In another pit [SU 0235 0125], mottled orange and grey silty clay was formerly worked.

Medium- to coarse-grained sand, close to the bottom of the underlying sand, was formerly worked in shallow pits [SU 0161 0106 and 0164 0125] near Deans Grove. A sample [SU 0283 0170] from north-west of Deans Grove consists of well-sorted, symmetrically skewed, fine-grained sand. CRB

In a track [SU 0712 9824] near Dudsbury, 1.2 m of white to pale grey, structureless, sandy clay rests on 2.45 m of pale buff, fine-grained sand with ferruginous sandstone ribs, overlying 1.2 m of obscured section, on 0.05 m of white clay, on 2.05 m of brown to buff, fine-grained sand. At Dudsbury, a 20 m-high cliff has good exposures [SU 0760 9785] in 1 m of grey sandy clay of the Broadstone Clay, above 5 m of fine- to coarse-grained, pale brown, cross-bedded, variably cemented sand with dewatering structures, resting on a prominent 0.35 m-thick bed of dark brown, hard sandstone, on 3 m of pale brown to brown, medium-grained sand and sandstone.

Laminated clays and fine-grained sands are exposed in the side of a track [SZ 1461 9473] on St Catherine's Hill, and were proved in a piston sampler hole [SZ 1460 9473] nearby. BJW

Outcrops between the River Stour and Poole Harbour

Mottled orange and grey silty clay and smooth, grey clay occur in the floor of the stream [SZ 0479 9650 to 0499 9663] which flows through Bearwood. Trial pits [around SZ 048 963] in the bottom of the old sand pit at Bearwood proved more than 2.7 m of stiff white and yellowish white, locally laminated, silty clay and clayey silt.

Boreholes [around SZ 0455 9640] south-east of Eastlands Farm proved over 5.3 m of firm to stiff, dark grey, laminated, silty clay with partings of fine-grained, pale grey sand; a thin bed of lignite was noted in one borehole. Mottled red and grey clay, close to the top of the Broadstone Clay, was seen in a temporary section [SZ 0397 9695] 650 m north-west of Eastlands Farm, and in a nearby ditch [SZ 0371 9702]. Boreholes [SZ 0404 9672; 0397 9672; 0389 9641] south-west of the last occurrence, proved up to 6.8 m of yellow and grey, locally laminated, silty clay, fine-grained sandy silt and stiff, grey silty clay (Freshney et al., 1985, p.48). Near Eastlands Farm, boreholes [around SZ 040 967] and auger holes [around SZ 042 967] proved up to 16.6 m of mottled pale grey and yellowish orange, dark grey to brownish grey clay; or stiff, locally laminated, fine-grained sandy and silty clay; or fine-grained sandy silt (Freshney et al., 1985, pp.44 and 47).

The sand beneath the Broadstone Clay is about 16 m thick south-west of Knighton Farm [around SZ 0465 9720], and is described in borehole logs as silty and fine to medium grained. Samples from boreholes 200 to 400 m to the west, range from very coarse grained to pebbly; the sorting varies from moderate to poor. Bluish grey silty clay, finely laminated, lignitic, very clayey, fine-grained sand and fine-grained sandy clay were observed in the bottom of the large, partially filled sand and gravel pit [SZ 030 970] north of Canford Heath. North of the pit, the underlying sand is fine grained.

Poor sections in mottled orange and grey silty clay, and pale grey, silty, fine-grained sand, occur in old brick pits [SZ 0265 9720; 0210 9685] west of the sand and gravel pit. In a third pit [SZ 0150 9690], 3 m of dark carbonaceous clay were recorded (White, 1917), but now only 0.4 m of mottled orange and grey silty clay, beneath orange, coarse-grained sand may be seen.

Road cuttings [SZ 0163 9740 to 0165 9762] south-west of Merley mainly expose mottled orange and grey silty clay, but beds of dark brown, lignitic clay up to 1 m thick, and fine-grained, well-sorted sand, and coarse-grained sand up to 2 m thick occur locally.

Dark brown, lignitic clay was formerly exposed to a depth of 4.5 m in the old railway cutting [SZ 0141 9772] south-east of Merley House (Reid, MS map, 1894). The basal Broadstone Clay was penetrated in a borehole north of the cutting [SZ 0139 9792], and in another at Oakley [SZ 0194 9837]. In the first, beneath 2.8 m of river terrace deposits, 3.6 m of yellowish brown sandy clay rest on 1.1 m of fine-grained, moderately sorted sand with thin clay beds. Below this there are 5.7 m of very sandy clay, interlaminated with fine- to medium-grained clayey sand; this lowest unit is regarded as part of the sand beneath the Broadstone Clay (Freshney et al.,

1985, p.49). The underlying sand in a third borehole [SZ 0098 9823], south-east of Merley House, was thicker and consisted of 12.7 m of medium- to coarse-grained, well-sorted sand, above London Clay.

Reid (MS map, 1894) noted 9 m of 'grey loam' in the railway cutting [SZ 000 979] at Ashington. Auger holes in the banks proved mottled orange and grey, pinkish and pale to dark greyish brown silty clay, and orange clayey silt; interbeds of coarse-grained orange sand up to at least 1.2 m thick also occur. The dinoflagellate cyst *Kisselovia coleothrypta* was obtained from a sample of dark greyish brown silty clay.

Medium grey, greyish brown, and mottled orange and grey clay occurs at several localities [around SZ 006 971, 0045 9683 and 0053 9670] on Broadstone Golf Course. The underlying yellowish brown, medium- to coarse-grained sand is exposed in an old pit [SZ 0063 9702], and floors the stream bed [SZ 015 971] and ditches [SZ 0085 9700; 0123 9700] to the east.

At Broadstone Post Office [SZ 0048 9583], some 3.2 m of 'clay' is reputed to have been encountered in a temporary exposure. The railway cutting [SZ 0032 9574 to 0018 9533] south of Broadstone has poor exposures in fine-, medium-and coarse-grained sands. A sample of the coarse-grained sand was moderately sorted; the fine- and medium-grained sands were poorly sorted. The following section [SZ 0025 9541] is exposed at the top of the cutting: pale grey, silty clay, 1.2 m, on yellow, fine-grained sand, 0.3 m, overlying buff-yellow, fine- to medium-grained sand, 0.3 m. Coarse- to very coarse-grained yellow sand is present at the bottom of the cutting. Several boreholes in York Road [around SZ 0035 9535], Broadstone, proved up to 4.3 m [at SZ 0034 9540] of yellow and grey clay and silty clay above sand. Reid (MS map, 1894) noted 9 m of sand with masses of white clay, laminated carbonaceous loam, lignite, and ironstone nodules with leaves in the now almost filled Broadstone brick pit [SZ 0085 9565]. Dark grey, and mottled orange and grey, silty clay occurs in its banks.

A small outlier [SY 975 962] of coarse-grained, cross-bedded, sand, up to 5 m thick, caps Cherrett Clump. An old pit has a number of degraded sections; one such section [SY 9754 9614] shows about 2 m of thinly bedded, cross-bedded, coarse-to very coarse-grained sand with scattered clay clasts. Cross-bedding measurements from the sand are included in Figure 14.

The Notting Hill–Henbury Plantation outlier extends in a north-easterly direction for about 1.25 km. Much of its outcrop is occupied by the Henbury sand and gravel pit. The maximum thickness of the Poole Formation is about 12 m. Cross-bedding measurements from the sands are incorporated in Figure 14.

A section [SY 9631 9732] on the south side of the pit showed:

	Thickness m
RIVER TERRACE DEPOSITS (THIRTEENTH)	
Gravel, generally 1 to 1.5 m thick, but locally up to	2.5
Poole Formation	
Sand beneath the Broadstone Clay	
Sand, coarse- to very coarse-grained, cross-bedded, orange, scattered small (up to 5 mm) lignite pieces	6.0
Sand, buff, fine-grained	1.0
Unexposed	c.3.0
London Clay	
Clay, very fine-grained, sandy, bluish grey	

CRB

Sections in the Broadstone Clay in the Beacon Hill pit [SY 982 951] are described in the Parkstone Clay section.

The junction of the Broadstone Clay and underlying sand can still be seen in the old pit [SY 9860 9653)] south of Wyatts Lane, Corfe Mullen. There, a section [SY 9856 9653] showed:

	Thickness m
Poole Formation	
BROADSTONE CLAY	
Clay, silty, pale grey	0.9
SAND BENEATH THE BROADSTONE CLAY	
Sandstone, ferruginously cemented, medium-grained	0.1
Sand, fine- to medium-grained, orange	3.0

The total thickness of the sand hereabouts is about 12 m.

Up to 3.8 m of pale grey, red, red and grey mottled, and brown and grey clay were proved in boreholes at Broadstone Middle School [around SZ 0115 9605]. Boreholes north of Poole Grammar School proved up to 3 m of stiff, pale grey clay, with some interbedded fine-grained sand and silt [around SZ 0165 9540].

In a road bank [SZ 0178 9529] on Canford Heath, 2 m of medium-grained, orange, cross-bedded, friable sandstone, in units 8 cm to 20 cm thick with a 2 mm-thick lateritic top, are overlain by about 10 m of mottled orange and grey silty clay. The contact between the two units dips 18° at N340°. Another road bank [SZ 0168 9444] revealed about 2 m of coarse- to very coarse-grained, cross-bedded, friable sandstone. Dark grey, silty clay with lignite was seen to a depth of 1.5 m [SZ 0208 9508] 900 m to the north-east. Boreholes and ditch sections [around SZ 033 945] on the south side of Canford Heath, proved up to 3.2 m of firm, bluish grey clay, red and orange mottled, laminated clay, yellowish brown, white, pale grey and brown, slightly sandy silty clay, and pale grey, fine-grained sandy silt (Freshney et al., 1985, p.53). CRB

Boreholes for Canford Heath shopping centre [around SZ 026 936] proved up to 1.8 m of firm, reddish brown, silty clay. Excavations for the Oakdale underpass revealed the following section at the junction of the Broadstone Clay and the underlying sand [SZ 0218 9324]:

	Thickness m
Poole Formation	
BROADSTONE CLAY	
Clay, reddish brown, sandy and silty, with silty sand partings; layers of sand near the base range up to medium-grained	1.5
SAND BENEATH THE BROADSTONE CLAY	
Sand, coarse-grained to pebbly, cross-bedded, grey; white, rotten flint grains common; some lamination picked out by pebble bands and thin carbonaceous layers	4.0

The lower part of the sand, and its junction with the Oakdale Clay, was exposed in a road cutting [SZ 0215 9331] 80 m to the north:

	Thickness m
Poole Formation	
SAND BENEATH THE BROADSTONE CLAY	
Sand, coarse-grained, orange to dark brown, with some fine gravel; well-banded and laminated	1.1
Sand, clayey, coarse- to very coarse-grained, orange	0.2
Sand, clayey, fine-grained, yellowish orange, with streaky lamination; some blocky clasts of pale grey, extremely sandy clay to clayey fine-grained sand; carbonaceous layers about 1 m down; grain size becomes medium, and the colour becomes brown downwards; strong marcasite stain near base; water weeping from sharp base	2.0

OAKDALE CLAY	
Clay, sticky, soft, pale grey, with carbonaceous material and ?rootlets; local masses of black to dark reddish brown, highly lignitic clay 0.3 m across; some siderite nodules	2.0

A borehole [SZ 0217 9323] at the subway proved 6.7 m of medium- to coarse-grained sand, beneath up to 1.9 m of stiff greyish brown mottled, fissured, silty clay [SZ 0225 9324]. In a borehole [SZ 0250 9303] in Valley Road, Oakdale, over 2.16 m of mottled pale greyish pink clay, with thin beds of fine-grained sand and silt, and becoming organic downwards, were proved. Brown silty clay occurs in the western part of Oakdale Cemetery [SZ 0240 9272]. CRB, ECF

Outcrops west and south of Poole Harbour

On the north side of Black Hill, there is a 2 m exposure [SY 9446 9124] of coarse- to very coarse-grained, cross-bedded, friable sandstone, in sets up to 15 cm high, capped by pale grey silty clay. Brown [SY 9447 9098] and pinkish grey [SY 9458 9085; 9463 9085] clays occur locally. Some 300 m to the east, road cuttings [SY 9476 9124; 9482 9139] reveal up to 2 m of fine-grained, well-sorted, symmetrically skewed sand, and medium-grained, moderately sorted, very positively skewed sand. Coarse-grained sand and grit was noted at a third locality [SY 9476 9134] hereabouts.

On Slepe Heath, the dominant lithology is coarse- to very coarse-grained or gritty, commonly ferruginous, locally cross-bedded sand [e.g. SY 9420 8508]. Local developments of pale grey and red-stained silty clay [around SY 942 855] on the southern part of Slepe Heath, and mottled orange and grey silty clay low down on the north side of Hartland Moor [around SY 9540 8567], may be the Broadstone Clay.

West of Arne, medium- to very coarse-grained, poorly sorted sandstone is exposed in low cliffs [SY 9593 8800; 9598 8829]. Some 150 m north-north-east, a lens of brownish grey, laminated silty clay, which dies out northwards, occurs within the sand [SY 9606 8846]. Northwards, 1 m of cryoturbated sand and gravel overlies 4 m of coarse-grained buff sand [SY 9609 8854]. A further 1.2 km north-east [SY 9705 8926], there are 4 m-high cliff sections where 1 m of medium-grained sand at the base is succeeded by 3 m of very coarse-grained sand and grit, which fines upward. It was from cliff sections hereabouts that Chandler (1961) obtained a rich flora. She recorded coarse-grained sand resting on mottled red and white pipeclay; the upper part of the cliff consisted of barren, coarse-grained sand and grit, below which were a few fossiliferous carbonaceous seams. North of Chandler's section, there are up to 8 m of yellow, medium- to coarse-grained sand with a coarse-grained ferruginous grit at the top [SY 9714 8942].

The Broadstone Clay is up to 12 m thick at Arne [around SY 970 884] and consists of pale to medium and, in places, dark grey, mottled orange, locally lignitic clay.

At Shipstal Point, there are 6 m-high cliff sections in which river terrace deposits overlie coarse- to very coarse-grained, cross-bedded, friable orange sand. In the middle of the northernmost section hereabouts [SY 9834 8838], there is a 0.3 m bed of pale grey silty clay. CRB

On Long Island the following section [SY 9864 8809] occurs:

	Thickness m
Soil	
Buff sand with scattered flints	c.1.0
River Terrace Deposits	
Sand, cross-bedded, and sand with gravel; cryoturbated; channelled base	up to 1.0
Poole Formation	
SAND BENEATH THE BROADSTONE CLAY	
Sand, coarse- to very coarse-grained, with two gritty beds; cross-bedded on a decimetre scale	1.8 to 2.0
Sand, clayey, very coarse-grained, gritty; thickens southward into a channel trending c.050°; clay clasts and clay rafts near base up to 0.8 by 0.4 m size	0.1–0.9
?OAKDALE CLAY	
Clay, very silty, grey, with some carbonaceous material and scattered lignite debris; laminated in places; gritty sand within clay about 0.2 m thick and 0.3 m above the base	2.3
Sand, coarse-grained (occurs below beach level)	0.2

A section on Round Island [SY 9870 8753] showed:

	Thickness m
Poole Formation	
SAND BENEATH THE BROADSTONE CLAY	
Sand, buff to orange, medium- to coarse-grained, cross-bedded, with some convolute bedding; variably ferruginously cemented	1.2
Sand, medium- to coarse-grained near top, but very coarse-grained, with coarse grit bands in lower part; clay clasts scattered throughout, but more common in basal part; sigmoidal cross-bedding common on 3–12 cm scale	1.8

Samples of the sand vary from medium to coarse grained and are well sorted and symmetrically skewed. A section [SY 9894 8741] on the east of the island showed up to 2 m of laminated, greyish brown, clayey silt, overlying 0.3 m of very fine-grained buff sand. Laminated silt also occurs at beach level [SY 9896 8747] farther north, and overlies fine-grained carbonaceous sand nearby [SY 9895 8749]. On the north of the island, medium- to coarse-grained sand is exposed [SY 9892 8753; 9883 8755]. CRB, ECF

An auger hole [SY 9859 8513] on Rempstone Heath proved:

	Thickness m	*Depth* m
Poole Formation		
SAND BENEATH THE BROADSTONE CLAY		
Sand, clayey, fine-grained	0.80	0.80
Sand, clayey, medium- to very coarse-grained, yellow	0.20	1.00
Silt, clayey, greyish brown	0.10	1.10
Sand, medium-grained, clayey, yellow	0.80	1.90
Silt, greyish brown, clayey	0.05	1.95
Sand, clayey, very coarse-grained, yellow; very silty clay beds up to 2.1 m thick; very gritty and almost conglomeratic at 2.1 to 2.8 m; very coarse-grained sand below to base	1.35	3.30

On the Goathorn Peninsula, a section [SZ 0093 8592] showed about 8 m of mainly fine- to medium-grained, cross-bedded sand, with a 0.2 m-thick bed of very coarse-grained gritty sand about 2 m from the top.

A second section [SZ 0130 8726] showed:

	Thickness m
4th River Terrace Deposits	
Sand, grey, gravelly	0.8
Gravel, orange, sandy	0 to 0.4

Poole Formation	
SAND BENEATH THE BROADSTONE CLAY	
Sand, medium- to coarse-grained, becoming coarser down to a grit at base; some clay clasts	3.0
Clay, grey, silty, structureless	0.3
Sand, fine-grained, comprising eight c.0.1 – 0.15 m fining-upward microcycles, each becoming clay-rich in top half	1.0

On the south side of Brand's Bay [SZ 0235 8556], yellow, and some red-stained, cross-bedded sand is exposed. The cross-bedding sets are 10 – 30 cm thick, and commonly planar.

An auger hole on Newton Heath [SZ 0007 8502] proved:

	Thickness m
Head	
Sand, medium-grained, with flints	0.8
Sand, yellow, very clayey, with scattered flints	0.3
Poole Formation	
SAND BENEATH THE BROADSTONE CLAY	
Sand, grey, very clayey, becoming a very sandy clay below 1.5 m	0.5
CLAY IN POOLE FORMATION	
Clay, grey with some red staining, very sandy, rapidly becoming pale brown and less sandy with depth; scattered brown clayey sand beds near top; passes down into brown silty, carbonaceous clay, with some patches of lignitic clay; traces of lamination	1.7

A second auger hole 30 m to the west proved only sand, some coarse-grained, some clayey and silty, to a depth of 3.1 m, suggesting that the clay in the previous section occupies a steep-sided channel. ECF

PARKSTONE CLAY AND UNDERLYING SAND

Outcrops north of the River Stour

North of Uddens Water, the sand beneath the Parkstone Clay is dominantly coarse to very coarse grained. A section [SU 0462 0361] in the road cutting near White Sheet Plantation revealed 0.5 m of cross-bedded, orange, coarse- to very coarse-grained sand, overlying 0.8 m of coarse- to very coarse-grained, orange-buff sand in thin beds, up to 3 cm thick. CRB

There are many old clay pits in the Ferndown area (Freshney et al., 1985, p.60). A borehole [SU 0881 0164] near Trickett's Cross proved 9.5 m of grey silty clay, beneath 2.8 m of made ground and drift. Another borehole [SU 0814 0173] penetrated 3.2 m of alluvium, on 5.6 m of laminated clay, on 1.2 m of sand. A further borehole nearby [SU 0804 0184] proved 15.5 m of fine- and medium-grained sand with thin clay partings beneath 4.5 m of drift.

Between West Parley and the River Stour, several boreholes proved clay; a typical sequence [SZ 0863 9744] is 3.3 m of river terrace deposits on 3.3 m of laminated silty clay. The BGS Parley Court No. 1 Borehole [SZ 0928 9649] proved 9 m of drift and Branksome Sand, on 14.4 m of dark greyish brown clay with wisps and laminae of fine-grained sand and silt, on 17.6 m of coarse- and medium-grained sand with clay bands. Dinoflagellate cysts from the Parkstone Clay in the Parley Court No.1 and 2 boreholes included *Areosphaeridium dictyostilum Kisselovia coleothrypta* and *Wetzeliella horrida* which indicate the *coleothrypta* Zone and a nearshore environment of deposition. BJW

Outcrops between the River Stour and Poole Harbour

The Parkstone Clay has been cut out beneath the Branksome Sand in the old sand pit [SZ 047 960] at Bearwood. A typical section [SZ 0487 9613] in the sand below the Parkstone Clay is as follows:

	Thickness m
Head	
Gravel, clayey, cryoturbated	1.5
River Terrace Deposits (Eighth)	
Sand and gravel, bedded, base channelled into underlying strata	3.0
Poole Formation	
SAND BENEATH THE PARKSTONE CLAY	
Sand, coarse- to very coarse-grained, cross-bedded in sets up to 5 m thick; stained orange where overlain by gravel, otherwise off-white; northwards this unit passes under clayey fine-grained sand and fine-grained sandy clay; bedding at top destroyed by dewatering structures	8.0

In a pit [SZ 0322 9640] west of Eastlands Farm, up to 2 m of ferruginous gravel rests in pockets on 6 m of coarse- to very coarse-grained, cross-bedded sand, in units 0.1 to 0.4 m thick. Bedding is picked out by a weak ferruginous cement, and some microfaulting is present. Excellent sections occur in the sand pit [SZ 030 968] to the north-west. In general, up to 3 m of river terrace deposits overlie 2 to 3 m of very coarse-grained, orange, thinly bedded, cross-bedded sand, above 6 to 8 m of white or yellow, fine- to coarse-grained or pebbly sand, with thin beds of pale grey or off-white clay and fine-grained sandy clay, resting on laminated, carbonaceous, clayey, fine-grained sand of the Broadstone Clay. The lithologies change rapidly across the pit, and no single section includes all the above features. Some beds are graded, and fine upward from fine gravel to a medium-grained sand. The grains and pebbles of the coarser beds consist mostly of subangular to subrounded flint, with a few of well-rounded quartz and patinated flint. Clay clasts up to 3 cm across are present in some sand beds. Elsewhere [SZ 0303 9673], there are impersistent clay breccias, up to 1.5 m thick, with clasts up to 10 cm across.

Some 800 m south-west of the pit, a borehole [SZ 0223 9630] proved 1.6 m of clayey gravel, on 2.8 m of stiff, pale grey clayey silt, at the base of the Parkstone Clay, on 25.6 m of fine-, medium- and coarse-grained sand.

Up to 26 m of sand are exposed in road cuttings [SZ 0162 9610 to 0197 9585] on the north-west side of Canford Heath. The sands lie in a north – south-trending syncline, with an 8° dip on the western limb and a 10° dip on the eastern limb. The lithologies vary, but the general sequence comprises thinly bedded (locally cross-bedded) orange, medium- to very coarse-grained sand up to 10 m thick, overlying up to 8 m of thinly bedded (locally cross- bedded), white, fine-grained, sand with thin (up to 0.4 m) pale grey silty clay beds, resting on more than 8 m of fine-, medium- and coarse-grained, yellowish orange sand. Up to 3 m of thinly planar-bedded (beds 0.02 to 0.10 m thick) sand of the uppermost unit are seen in a pit [SZ 0168 9620] north of the cutting; greyish white silty clay occurs in the bottom of this pit.

Sections in Parkstone Clay in a track [SZ 0222 9569; 0225 9542] show 0.4 m of pinkish grey silty clay resting on off-white, fine-grained sand. Cuttings to the south expose a variable sequence of sands at a lower stratigraphical level. One section [SZ 0224 9527] shows 1 m of laminated, pale grey, silty clay in beds up to 0.1 m thick, overlying more than 4 m of coarse-grained, buff-brown, cross-bedded sand. Mottled grey and medium brown clay occurs

[SZ 0246 9525] beneath gravel 200 m east of the road cutting [SZ 0246 9525]. Farther east [around SZ 030 953], Parkstone Clay is locally cut out beneath Branksome Sand. Between 2.4 and 8.4 m of stiff, laminated, very silty clay and clayey silt were proved in boreholes [around SZ 0375 9508] on the east of the Heath. The feather edge of the Parkstone Clay, beneath 2.1 m of river terrace deposits, is represented by 0.9 m of stiff, yellow, silty clay in a borehole [SZ 0385 9536] to the north. CRB

North-east of Alderney hospital, Parkstone Clay consists of brown and grey silty clay or sandy silt; clay was proved to depths of 8.4 and 14 m in two boreholes [SZ 0456 9466; 0466 9464] hereabouts. Farther north [around SZ 043 948], Branksome Sand is channelled into Parkstone Clay, and the clay may locally be cut out. The western margin of a south-west-trending channel cut into the clay occurs in pit [SZ 0408 9480] north-west of the hospital. A section [SZ 0399 9477] in the western part of the pit showed:

	Thickness m
Poole Formation	
PARKSTONE CLAY	
Clay, white, structureless, cryoturbated, with gravel lobes at top, and tabular rafts of sand	1.2
Clay, pale grey, in seams 2 to 5 cm thick, interbedded with fine-grained sand; below 0.5 m from top, the clay seams are as thin as 3 mm and are disrupted to flat pebble clasts; the sand content increases in lower part of unit	1.1
Clay, brown, roughly laminated with fine silt or sand partings; some 1 cm beds and lenticles of sand	1.5

Some 250 m north-west of the pit, 3 m of cross-bedded, coarse- to very coarse-grained sand, slightly pebbly in the lower part, were seen; a lens of pale grey silty clay and clayey silt, up to 0.3 m thick, occurs 0.5 m above the base. A borehole [SZ 0388 9477] farther west, proved 11 m of stiff, dark brown clay with sand laminae. Boreholes in the old brick and tile pit [SZ 040 946] encountered stiff, brownish grey, laminated, silty clay beneath fill. The disused Manning's Heath Pit [SZ 037 943] has up to 10 m of fill, above stiff, pale and chocolate-brown mottled, fissured, very silty clay.

A pit [SZ 034 937] to the south-west contains up to 4.6 m of fill, above up to 3.9 m of hard, dark grey or brown, laminated clay. Water wells in the pit, not accurately sited, proved up to 18.9 m of 'hard blue clay', or 'hard mottled clay and clayey rubble' above sand, according to drillers. Sections along the road north-west of the pit are in medium- to very coarse-grained, cross-bedded, locally laminated, orange sand; some dewatering structures were noted [SZ 0326 9382].

A former pit [SZ 0325 9275] south of the old Wareham Road is now filled, but boreholes penetrated up to 7.3 m of fill overlying up to 7.6 m of stiff to very stiff, pale grey, laminated, carbonaceous silty clay, resting on fine-grained sand. A section [SZ 030 926] in the Kinson Pottery Pit formerly showed 3 to 4.5 m of red-mottled white clay (White, 1917). The top of the clay, beneath 15 m of Branksome Sand, was planar and 'brightly reddened'.

In a pit 550 m south-west of Oakdale Cemetery, the following section [SZ 0238 9216] is exposed:

	Thickness m
Poole Formation	
PARKSTONE CLAY	
Clay, silty, mottled orange and grey	1.0
Clay, silty, mottled red and grey	0.5
SAND BENEATH THE PARKSTONE CLAY	
Sand, coarse- to very coarse-grained, trough cross-bedded, with some dewatering structures; top 10 cm ferruginously cemented	4.0

The contact between the clay and sand dips 10° NE.

A section [SZ 0276 9215] south of Oakdale Cemetery exposes 0.1 m of pale grey silty clay, resting on 2 m of coarse- to very coarse-grained, cross-bedded sand with thin pipeclays. Boreholes [c. SZ 0297 9237] north of the cemetery proved up to 3.3 m of grey silty clay with some thin interbeds of sand, beneath Branksome Sand. At least 3 m of homogeneous, medium grey, silty clay overlie fine- to medium-grained, poorly sorted sand on a building site [SZ 031 915] west of Parkstone Station; the base of the clay dips 5°E. Boreholes at Parkstone Telephone Exchange [around SZ 0370 9165] proved up to 10.3 m of firm to stiff, mottled grey, red, purple and yellow, fissured silty clay, beneath 4.9 m of made ground and head.

Baden Powell Middle School [SZ 0365 9125], Parkstone, is built in the bottom of an old, partially filled clay pit. There, up to 3.7 m of stiff pinkish grey, dark and pale grey, and pale brown mottled silty clay, and pale grey clayey silt were proved. Boreholes near the railway [around SZ 0365 9140] proved more than 3 m of very stiff, laminated, greyish brown clay.

A pit [SZ 0365 9097] at Parkstone exposed a sand within the Parkstone Clay. The lowest 3.5 m of this sand and 15.2 m of the underlying clay were proved in the BGS Homark Borehole [SZ 0364 9096]. The composite section is as follows:

	Thickness m	*Depth* m
MADE GROUND		
Sand and gravel, clayey, organic	2.0	2.0
Poole Formation		
PARKSTONE CLAY		
Sand, black, slightly clayey, medium- to fine-grained, waterlogged	1.5	3.5
Sand, very fine-grained, and silt, with much fine- to medium-grained sand below 5 m; dark greyish brown, clayey; some fine lignitic debris	2.0	5.5
Clay, medium grey, with some marcasite staining and listric surfaces (possible palaeosol)	8.0	13.5
Sand	0.5	14.0
Clay, grey to greyish brown and brown, streaky, plastic, sticky,with some marcasite staining; darker brown and silty at 18 to 18.5 m; dominantly greyish brown, plastic; below 18.5 m only a small silt content	6.7	20.7

A section in the north face of a pit [SZ 038 910], east of the Homark Borehole, shows 1.3 m of Branksome Sand with a channelled base, resting on 1 m of dark brown, lignitic clay, interlaminated with fine-grained sand; the clay laminae have sharp bases and pass up into the sand laminae. Non-age-diagnostic dinoflagellate cysts from the clay are indicative of a nearshore environment. An old sandpit [SZ 0418 9055], 700 m south-east of the Homark Borehole, formerly exposed 6 m of sand with seams of 'loam', above 4.5 m of carbonaceous loam and lignite (Reid, MS., map 1894).

Boreholes [around SZ 0350 9085] near the old Parkstone brickworks proved up to 3.5 m of mottled red and grey, or greyish brown, firm, silty clay and clayey, fine-grained sandy silt. A well [SZ 0342 9064] 250 m south-west of the brickworks, proved 2.9 m of clay and sand, above 14.5 m of black and brown clay with a 0.9 m interbed of 'running' sand, which in turn rests on sand (thickness unproven).

The junction between the Parkstone Clay and the underlying sand was seen at two places [SZ 0289 9071; 0280 9073] in the old cliffs north of Parkstone Bay. The section at the first locality shows:

	Thickness m
Poole Formation	
PARKSTONE CLAY	
Sand, very fine-grained, silty, finely laminated, lignitic, with a fine- to very fine-grained lignitic sand at the base	0.8
Clay, silty, pale grey, with rootlets; lateritic base	1.8
SAND BENEATH THE PARKSTONE CLAY	
Sand, fine-grained, thinly bedded, orange-buff	0.8
Sand, fine-grained, massive, buff	0.8

The highest sand is a bed within the clay. In the second section, pods of fine-grained homogeneous sand, up to 0.7 m thick and 20 m long, occur within 2.3 m of pale grey and mottled red and grey silty clays. Carbonaceous 'loam' was formerly exposed in the railway cutting [SZ 0277 9082]. Red-stained, pale grey silty clay is poorly exposed in the cliffs farther east [SZ 0315 9073]. Poor sections [SZ 0317 9000 to 0325 8924] in the cliffs to the south-east, show brown sandy clay with silt and very fine-grained sand layers at beach level.

Reid (MS. map, 1894) noted a section [SZ 0406 8931] in the Lilliput cliffs which consisted of 2.4 m of sand, above 3.6 m of 'clay loam'. A borehole 1.25 km to the south-east [SZ 0499 8849] proved, beneath 3.4 m of sand and gravel, a total of 10.8 m of dark grey and brown, laminated, silty clay and dark grey, laminated, clayey silt with a 1 m interbed of medium-grained sand, resting on clayey sand.

At the base of the cliffs at Poole Head [SZ 0520 8854; 0531 8865], 2.4 m of laminated 'loam' and carbonaceous clays, and laminated blue carbonaceous 'loam' with '*Teredo*'-bored lignite, were formerly exposed (Reid, MS, BGS). In the cliff section at Sandbanks [c.052 886], Chandler (1963) recorded, beneath 'barren' sands, a richly fossiliferous lignitic lenticle 3.6 m long and 0.6 m thick, composed of a matted mass of twigs, wood, fruit and seeds set in a sandy matrix. This rested on 1.5 m of dull purplish, laminated silt with thin carbonaceous seams.

In the head-filled valley [SZ 043 928] south-east of Newtown, 1.6 m of firm, dark grey silty clay is overlain by 2.4 m of gravel and gravelly sand.

Beacon Hill

The Beacon Hill clay pit [SY 982 951] and the Corfe Mullen clay pit [SY 980 952] expose excellent sections in the upper part of the Poole Formation. For convenience, the whole Poole Formation sequence exposed in these pits is described in this section. The Broadstone Clay in the Corfe Mullen pit is not well exposed; the general sequence of clay and sand comprises:

	Thickness m
Poole Formation	
BROADSTONE CLAY	
Clay, mottled dark grey and red	c.4 to 5.0
Clay, pale grey	c.2.0
SAND BENEATH THE BROADSTONE CLAY	
Sand, fine-grained, thinly bedded and finely cross-bedded, and cross cutting; bedding units vary between 0.1 and 0.4 m thick	10.0
Sand, fine- to coarse-grained, thinly bedded and cross-bedded; cross beds commonly separated by thinly bedded (up to 8 cm thick) silty sand; some alternating clay and sand beds up to 0.6 m thick	3.0
Sand, coarse-grained to very coarse-grained, cross-bedded, with common scattered clay clasts — mostly less than 1 cm, but some up to 15 cm across. The coarser clasts occur in the base of the unit and become finer upwards; finer clasts follow the cross bedding	15.0

Cross-bedding measurements from the sands are included in Figure 14. A higher section [SY 9787 9502] exposed the junction between the sand beneath the Parkstone Clay and the Broadstone Clay:

	Thickness m
Topsoil	0.3
Head	
Clay, pebbly, mottled orange and grey	0.5 to 1.3
Silt, clayey, with scattered flints; some clayey coarse-grained sand, mottled orange and grey	0.4 to 1.0
Poole Formation	
SAND BENEATH THE PARKSTONE CLAY	
Clay, silty, pale grey and orange	0.4
Sand, medium- to coarse-grained, cross-bedded, buff	0.3 to 2.0
BROADSTONE CLAY	
Clay, silty, mottled red and pale grey	0.3

In another section close by, the sand beneath the Parkstone Clay consists of 2 m of very coarse-grained and gritty, cross-bedded sand. The bedding is convoluted and there are some small-scale syndepositional faults.

A section [SY 9796 9500] near the base of the sand beneath the Parkstone Clay, showed up to 6 m of medium-to coarse-grained, cross-bedded friable sandstone. The cross-beds vary in thickness from units of 2 to 50 cm. Bedding in the upper part of the sequence is locally convoluted; tabular, postdepositional, cross-cutting, thin (up to 1 cm) ferruginous sandstone layers occur at intervals of 0.1 to 0.4 m (Plate 4). Scattered clay clasts and lignite fragments up to 1 to 2 mm thick and 15 mm long occur throughout; some of the cross beds have clay bases.

In the floor of the pit, the top part of the Broadstone Clay is exposed [SY 982 952]. There, pale and dark grey clay with a lateritic cap is overlain by coarse- to very coarse-grained and gritty, cross-bedded sand.

Two sections [SY 9826 9523; 9831 9509] to the east and south-east, at about the same stratigraphical level within the sand beneath the Parkstone Clay, reveal clay and sandstone breccias (Plate 3). At the former, there are five levels of breccia within a 6 m section. There, the beds are up to 0.8 m thick and 5 m across; clasts of sandstone are up to 0.1 m thick and 0.5 m across. A small syndepositional fault with a 0.7 m displacement was seen. The breccia at the latter locality has clasts up to 0.4 m thick and 1 m long; the breccias are interbedded with thinly bedded, fine- to medium-grained sand with clay-rich partings.

A clay seam, up to 2 m thick, in the middle of a face in the eastern part of the pit [SY 9835 9517] is a lenticular unit within the sand beneath the Parkstone Clay. At one place [SY 9834 9520], its base consists of a dark grey structureless clay resting on the ferruginously cemented top of a coarse-grained sandstone. The dinoflagellate cyst *Kisselovia* cf. *coleothrypta*, indicative of the *coleothrypta* Zone, was obtained from the clay.

Plate 3 Clay-clast breccia within sands of the Poole Formation, Beacon Hill Sand Pit. Clasts of poorly indurated silty mudstone vary from 1 cm to more than 0.3 m. The thickness of the clay clast beds varies from 0.6 to 1.2m.

Close by, another section [SY 9832 9524] showed 1 m of medium grey, thinly laminated, silty clay resting on 1.5 m of finely laminated, fine-grained sand, silty sand and friable sandstone, with thin (up to 2 cm) greyish brown clay beds. The top of the clay was exposed in a third section [SY 9835 9517]:

	Thickness m
Poole Formation	
SAND BENEATH THE PARKSTONE CLAY	
Sand, coarse-grained, cross-bedded, ferruginous	0.15
Sand, coarse-grained, thinly interbedded with olive-grey clay	0.18
Sand, coarse-grained, ferruginous, with thin clay partings	0.15
BROADSTONE CLAY	
Clay, medium grey, laminated	c.1.50
SAND BENEATH THE BROADSTONE CLAY	
Sandstone, medium-grained, ferruginous	1.20

A higher face, farther east in the pit [SY 9844 9520], exposes Parkstone Clay and the overlying Branksome Sand. The southern end of the section [SY 9840 9511] shows 1.9 m of clay, pale and medium grey in the lower part, with the top 0.4 m red-brown. Farther north in the face, a section [SY 9840 9524] in the Parkstone Clay comprises 0.9 m of mottled orange and grey silty clay, overlain by a ferruginously cemented coarse-grained, cross-cutting sandstone. A few metres west of this section, the Parkstone Clay is completely cut out beneath the Branksome Sand, but reappears 20 m to the west.

Outcrops in Poole Harbour

Parkstone Clay occurs on Brownsea Island, and the sand that crops out on Furzey and Green islands is probably the sand unit beneath it. Many sections on Brownsea Island expose the junction of the Parkstone Clay and the Branksome Sand. The most westerly [SZ 0104 8798] shows:

	Thickness m
River Terrace Deposits	
Gravel, sandy	1 to 2
BRANKSOME SAND	
Sand, medium- to very coarse-grained, tabular-bedded on a decimetre scale; about 3 m from top there is a 0.1 m-thick brown silty clay underlain by a grit band	c.5.0

Plate 4 Convolute bedding (?dewatering structures) in sands of the Poole Formation, Beacon Hill Sand Pit.

Poole Formation

PARKSTONE CLAY

Clay, brown, silty, slightly carbonaceous, structureless in part, but also with common lamination; impersistent sand layers ranging from 0.5 mm to several millimetres thick; scattered beds of fine-grained sand up to 8 cm thick also occur; laminations and thin lenses of fine-grained sand more common towards base; clay layers have sharply defined tops and bottoms	10.0

The dinoflagellate cyst *Kisselovia* cf. *coleothrypta* was recovered from the Parkstone Clay.

A second section [SZ 0215 8750] showed:

	Thickness m
Branksome Sand	
Sand, medium-grained at top, very coarse-grained and gritty with clay clasts at base; some decimetre-scale cross bedding; thin white pipe-clay seam 0.5 m from top	over 3.0
Poole Formation	
PARKSTONE CLAY	
Clay, silty, greyish brown, structureless	0.3
Sand, medium-grained, fining upward from very coarse grit at base	3.5
Clay, dark brown, laminated with very fine-grained sand laminae; sand layers become thicker towards the base (up to 3 cm)	1.7
Sand, very fine-grained, carbonaceous, with irregular pods of very fine-grained sand	0.3
Clay, dark brown, carbonaceous, structureless	over 1.8

Dinoflagellate cysts from the Parkstone Clay here include *Homotryblium tenuispinosum* and *Kisselovia* cf. *coleothrypta* indicative of the *coleothrypta* Zone and a lagoonal environment of deposition.

On the north side of Brownsea Island there are several infilled shafts [around SZ 0170 8854 and 0202 8845] at sea level, from which Parkstone Clay was formerly extracted.

On Furzey Island, the following section [SZ 0085 8694] was seen:

	Thickness m
River Terrace Deposits	
Sand, orange	up to 1.0
Poole Formation	
SAND BENEATH THE PARKSTONE CLAY	
Clay, silty, grey, orange-stained, silty, structureless	up to 1.0

Sand, fine-grained, finely bedded, with clay laminae up to 2 cm, but mainly 1 to 2 mm; more clay laminae towards the base, which is channelled into beds below	0 to 2.3
Clay, very sandy, pale brown, with some ferruginous material	1 to 1.7
Sand, medium- to very coarse-grained, with grit bands; strongly cross-bedded, with many planar avalanche foresets; on some foresets, the sand forms decimetre-scale fining-upward units; ferruginously cemented sand layer 0.4 m thick, about 0.5 from base, is cut into the bed below; convolute bedding in one place at top of cliff	5.0
Sand, very coarse-grained, with scattered clay clasts; clay-clast conglomerate locally at base; concave-upward, trough-type cross-bedding	0.8
Silt, grey, very clayey, structureless	0.2 to 1.0
Sand, grey, medium-grained	1.0

Another section [SZ 0115 8720] showed:

	Thickness m
Poole Formation	
Sand beneath the Parkstone Clay	
Clay, carbonaceous, brown, silty, interlaminated with fine-grained grey sand; laminae range from 1 to 10 mm, but are impersistent; tops and bottoms of the clays are sharply defined; some lignite fragments and marcasite nodules	0.7
Sand, fine- to medium-grained, with many brown clay laminae ranging in thickness from 2 to 5 mm; some cross-bedding on a decimetre scale, with some clay drapes on foresets	1.3
Sand, medium-grained, structureless	0.1
Sand, fine-grained, laminated, with thin clay layers	0.1
Sand, medium- to coarse-grained, cross-bedded	0.3
Sand, fine-grained, slightly carbonaceous, with clay laminae	0.3
Sand, medium-grained, structureless	2.0

A third section [SZ 0120 8719] on the north side of the island showed:

	Thickness m
Blown Sand	
Sand, yellow, medium-grained	0.3
Poole Formation	
Sand beneath the Parkstone Clay	
Clay, dark brown, carbonaceous, with sand laminae which become commoner towards the top; layers of lignitic debris; occupies a north-east-trending channel; some discordance within the channel, with some clay layers cut out by sandier beds in the upper part of sequence	max 2.0
Sand, medium- to coarse-grained, with clay laminae; laterite at top	0 to 2.0
Sand, medium- to very coarse-grained, with weakly ferruginously cemented layers	0.7

A section [SZ 0041 8670] on the north-west of Green Island showed:

	Thickness m
Poole Formation	
Sand beneath the Parkstone Clay	
Grit, quartzose, yellowish orange, with erosional base cutting down 0.3 m	0.15 to 0.3
Sand, fine-grained, buff to orange-brown, with a ferruginous cement at top and base; slightly erosional base; small-scale cross-lamination on a centimetre scale	0.3 to 0.6
Sand, fine-grained, ranging to very silty clay, arranged in four fining-upward microcycles varying in thickness from 0.2 to 0.4 m; the topmost cycle terminates in silt, the others in silty clay; top two cycles are cross-laminated in their lower parts	1.0
Sand, very fine-grained, in three fining-upward cycles, the top of each consisting of 4 to 10 cm of clayey fine-grained sand and silty clay; much cross lamination on a centimetre scale	1.3
Sand, fine-grained, in at least four fining-upward cycles of 0.15 to 0.2 m thickness; the microcycles appear to occupy a major east–west channel	3.0
Gap in exposure	0.5
Sand, medium-grained, with a little coarse-grained sand at base, cross-bedded on a decimetre scale	0.5

Another section nearby [SZ 0040 8665] shows an angular discordance at the base of the sand bed below the grit. At the top of the cliff, the grit band contains clay clasts, and cuts down 1.2 m into the microcycle unit. About 50 m south of the last locality, the erosional base of the grit and coarse-grained sand cuts down to within 1.5 m of beach level, and then rises steeply southwards; the channel margin trends east–west. About 50 m farther south, a structural dip of 16° at N 140° brings the grit down to beach level.

ECF, CRB

BRANKSOME SAND

Stratigraphy

The Branksome Sand, 70 m thick, is named after Branksome Chine [SZ 069 090], and crops out in a 5 km-wide arc from Ferndown in the north, through Parley, the northern part of Bournemouth and the Branksome area of Poole, to Brownsea Island in the south. It is well exposed in cliff sections from Bournemouth Pier westwards to Canford Cliffs, though cliff protection schemes are steadily reducing the amount of exposure (Figure 16).

The base of the formation is taken at a transgressive erosion surface at the top of the Parkstone Clay or its underlying sand, and is usually marked by an influx of coarse sediment. Locally, the basal beds are fine-grained, well-sorted, and commonly red-stained sands; in the Alderney area, they are channelled into the Parkstone Clay and are ferruginously cemented at their base.

The Branksome Sand Formation consists of eight fining-upward cycles, of which the upper seven were lettered A to G in ascending order by Plint (1983b) (Figure 8). Plint's seven cycles together constitute his transgressive cycle 3 (Figure 9). An idealised cycle commences with a very coarse-grained sand, often containing clay clasts, resting on an erosion sur-

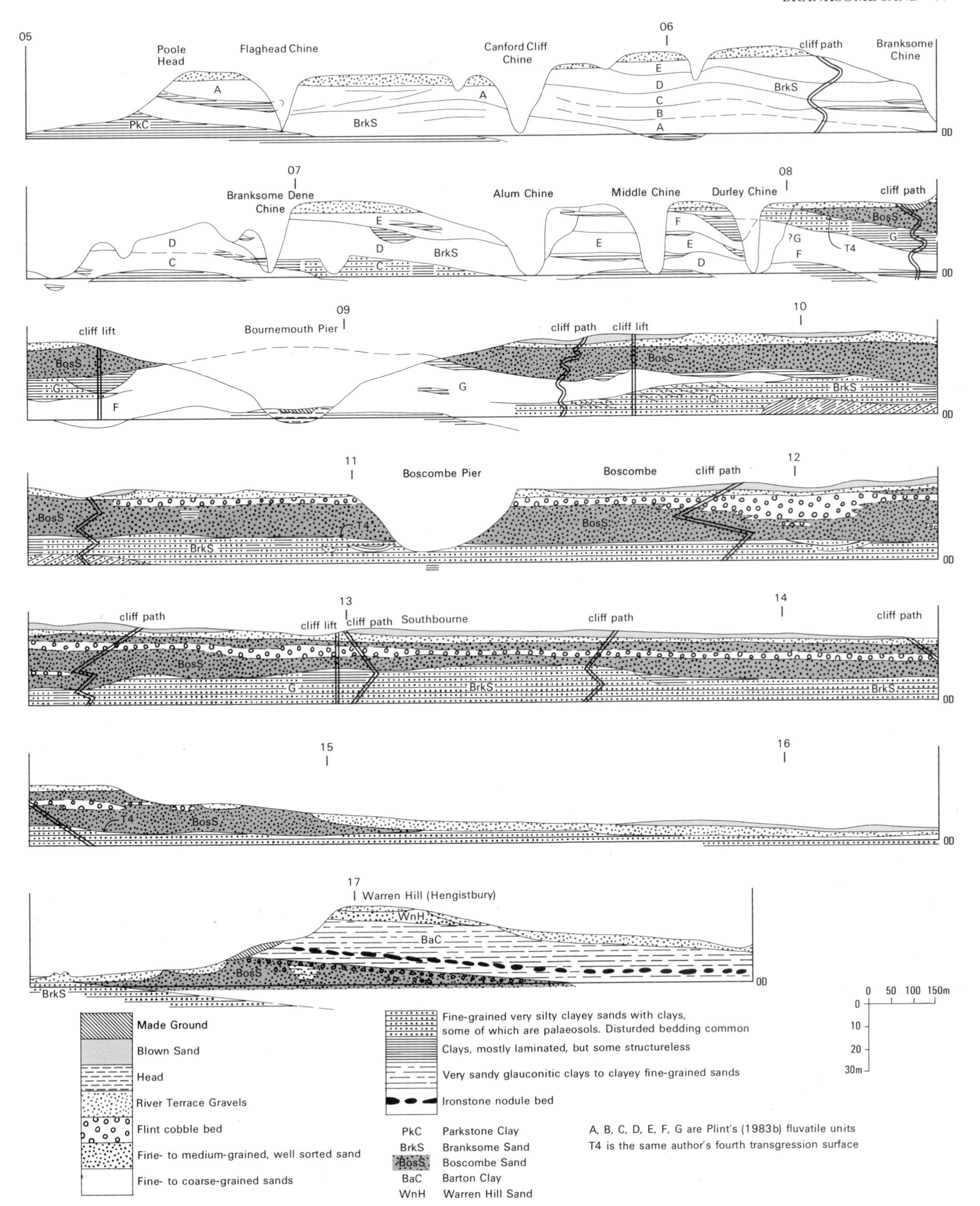

Figure 16 Cliff sections in the Poole Formation, Branksome Sand, Boscombe Sand and Barton Clay between Poole Harbour and Hengistbury Head (partly after Plint, (1988)).

face. This is succeeded by coarse- to medium-grained sand showing large-scale planar cross-stratification, which passes up into medium-grained sand in which the cross-bedding is on a decimetre scale. The cycle is capped by interbedded fine-grained sand and silty clay.

The sands are commonly strongly planar cross-bedded in sets up to 2 m thick; cross-bedding directions are very variable (Figure 14). In the Alderney area, several north–south-trending sand-filled channels, which cut into the top of the Parkstone Clay, have cross-bedding indicative of current flow from the north. Convolute bedding, common in the Branksome Sand, was probably formed by the rapid dewatering of saturated, poorly compacted sediment.

Mineralogically, the sand consists mainly of quartz, with some flint grains, which are more noticeable in the coarser-grained sands. The grains are mostly subangular to subrounded, but the very coarse grains are rather angular. The mean grain size ranges between coarse (0.33ø (800 μm)) and very fine (3.2ø (108 μm)), with an average of medium (1.98ø (253 μm)). A wide range of sorting is present from very well sorted (0.17ø) to poorly sorted (2.2ø), with a moderate average (0.65ø). The skewness also has a wide range (– 0.22 to 0.8), with an average of 0.23.

In some cycles, a structureless silty clay occurs at the top; in others, laminated carbonaceous, locally plant-rich clay, silt, and fine-grained sand occupy channels high in the cycle (Chandler, 1962; 1963). Palaeosols are present in some of the clays, but they have commonly been removed by erosion. The channels range from 90 to 170 m wide, and from 6 to 18 m deep in the Bournemouth cliffs (Plint, 1983b). The channels are most common at the top of Plint's cycles C, D, F and G; one occurs at the top of a poorly exposed unnamed cycle that underlies Cycle A between Poole Head and Flag Head Chine. The bedding within them is irregular, and slump structures are locally common. The tops and bottoms of the channel fills are well defined, but the margins are not.

One of the clay-rich units at the top of Cycle C (Freshney et al., 1985, p.91) shows great lateral variability in its sand content (Plate 5); it becomes increasingly sandy eastwards from Branksome Dene Chine to Alum Chine (Figure 16). The bedding also changes from centimetre-scale lamination at the western end to decimetre-scale and larger towards the

Plate 5 Cross-bedded sands of Cycle D overlying laminated, disturbed brown sandy clay and sand of a channel infill sequence at the top of Cycle C, Branksome Dene Chine.

east. The section shows much disturbed bedding and rotational slump structures close to the channel margins. Sands of Cycle D appear to have cut down into the sandy part of the cycle (Figure 16). Plint (1983a) interpreted the fine-grained sediments occupying the channel as fluviatile in origin.

The western end of the cliffs, between Poole Head and Branksome Dene Chine, exposes the lowest of the five fining-upward cycles (Figure 16).

Sediments east of Durley Chine, interpreted as belonging to Cycle G, are poorly exposed and may be a finer-grained, eastern equivalent of Cycle F. East of Bournemouth Pier, however, Cycle G consists of fine-grained sand, silt and clay (Figure 16), which pass eastwards into thin and rhythmically bedded, fine-grained clayey sands and silty clays with common carbonaceous and rootlet beds, and bidirectionally cross-bedded sands containing marine molluscs and *Ophiomorpha*, a trace fossil (Plint, 1983b). The above beds were called the Bournemouth Marine Beds by Gardner (1879). Ord (1914) reported lignite logs, some bored by *Teredo*, in the sands. A persistent laminated clay capped by a bleached palaeosol occurs at the top of Cycle G. This clay gives rise to a perched water table and a pronounced spring line in the cliffs.

Inland exposures are rare, but boreholes indicate clay bodies at similar stratigraphical levels to those exposed in the cliffs.

Biostratigraphy

Age-diagnostic dinoflagellate cysts were obtained from the Branksome Sand only in the Bournemouth cliff sections. Samples from ten inland localities (Table 1, localities 15 to 22) have sparse nondiagnostic microfloras. Those from the cliffs have been reported on by Costa et al. (1976). Additonal samples were collected by the authors and by Mr R V Melville from beds lower down in the sequence than those of Costa et al. The oldest dated BGS sample, from Canford Cliff (Table 1, locality 26), contained only a poor dinoflagellate cyst assemblage, but did include the index species of the *coleothrypta* Zone. About 400 m east-north-east of Eastcliff (Table 1, locality 24) a brown carbonaceous clay yielded *Cordosphaeridium funiculatum* and *Glaphyrocysta undulata*. The presence of *Glaphyrocysta intricata* and *G. vicina* in a sample from near the Toft Zigzag (Table 1, locality 25) indicates the *intricata* Assemblage Zone. The coastal section eastwards from this locality to Southbourne also yielded *intricata* Assemblage Zone floras (Costa et al., 1976).

The shelly marine fauna recorded by Gardner (1879) in Cycle G does not have any biostratigraphical significance. It includes species of the gastropods *Calyptraea, Cerithium, Natica* and *Phorus*, and the bivalves *Arca, Modiola, Ostrea* and *Tellina*.

Conditions of deposition

The Branksome Sand contains none of the widespread tidal flat or lagoonal back-barrier clays present in the Poole Formation; clays are of only limited extent and are restricted to channels. The formation, as seen in the cliffs at Bournemouth (Plint, 1983a), consists mainly of sands disposed in fining-upward cycles. Plint regarded these sands as multistoried point-bar sands, and the clays as fluviatile channel-fill deposits. However, these clays contain a sparse marine dinoflagellate cyst assemblage, which indicates some access to the sea; the sands show a more diverse origin.

At the base of the Branksome Sand, at several localities near Poole, there are a few metres of fine-grained, well-sorted sand which may represent a marine beach-barrier deposit which transgressed the intertidal and lagoonal clays of the Poole Formation. This event probably correlates with the basal Selsey Sand transgression of the adjacent Southampton district (Edwards and Freshney, 1987b). It is probable that a stratigraphically higher barrier sand, intermediate between the fluviatile sands in the Poole area (Plint's (1983a) multistoried point-bar sands) and marine shelf sediments to the east in the Southampton district, was present just to the east of the present district. This barrier sand may be represented in the Christchurch Borehole, where much of the Branksome Sand consists of coarsening-upward sequences of well-sorted sands with a fairly symmetrical grain-size distribution, and by the cross-bedded sands of Cycle G ('Bournemouth Marine Beds') at the top of the formation at Bournemouth. Plint (1983b) described this sequence in terms of fining-upward micro-cycles, and interpreted the beds as estuarine channel deposits.

Details

West Moors to Christchurch

An exposure [SZ 0943 9933] on Parley Common showed 3.5 m of fine- to coarse-grained, pale brownish grey sand, poorly cross- bedded, with some ferruginously cemented bands. A borehole [SZ 0965 9830] near East Parley proved, beneath 4.9 m of drift, 6 m of silty clay resting on 1.8 m of medium-grained sand

On the south flank of Lions Hill, a temporary section [SZ 1005 0381] revealed the following:

	Thickness m
Branksome Sand	
Sand, medium- to very coarse-grained, interlaminated with beds of pale grey clay up to 1 cm thick	0.90
Sand, coarse- to very coarse-grained, planar cross-bedded, with lenses of white silty clay	1.40
Sand, medium- to coarse-grained, thinly bedded, with white silty clay in irregular masses up to 2 cm across	1.15

On the south side of David's Hill, 2 m of coarse-grained grey sand with a 3 cm-thick mud-flake conglomerate at the base, are exposed in a roadside cutting [SU 1297 0355]. On Barnsfield Heath, an excavation [SU 1118 0042] revealed the following section:

	Thickness m
Branksome Sand	
Sand, very fine-grained, silty, cross-laminated; interbanded in up to 10 cm units with laminated, silty, carbonaceous clay	2.10
Sand, fine- and coarse-grained, with some clay content; dewatering structures common	1.90

Some 2.15 m of fine-grained, pale brown sand, with thin mauve clayey mudstone lenses up to 1 cm thick, were seen in a temporary exposure [SZ 1213 9987] south of Barnsfield Heath. A nearby stream section [SZ 1220 9962] revealed:

	Thickness m
Head	
Gravel, flinty	0.80
Branksome Sand	
Sand, fine-grained, pale brown, locally cross-bedded, with thin ferruginous layers and two thin seams of small subangular flints at 0.20 and 0.25 m below the top	1.70
Clay, greyish brown, silty and sandy, with coarse-grained sandy bands up to 1 cm thick	0.20
Clay, black, carbonaceous, plastic	0.05
Sand, fine-grained, brown, interlaminated with grey silty and sandy clay	1.10 seen

A borehole [SZ 1083 9947] in the drift-covered area near Hurn proved 4.1 m of medium-grained sand, pebbly in places, with thin silty clay laminae (Clarke, 1981, p.72). BJW

Several boreholes encountered Branksome Sand beneath up to 5.2 m of alluvium and river terrace deposits between Ringwood and Bisterne Manor (Clarke, 1981). One [SU 1538 0229] proved 4.4 m of clayey, yellowish orange, lignitic silt, on 0.8 m of laminated, silty, dark brown, lignitic clay, on 1.8 m of silty, dark brown, lignitic sand. ECF

A cutting [SZ 1314 9633] on the Christchurch–Hurn road reveals 5 m of fine-grained sand with ferruginous ribs. The A338 Bournemouth–Ringwood road cuts in the basal part of the Branksome Sand south-east of Hurn, and a good section [SZ 1357 9644] occurs at its northern end:

	Thickness m
Branksome Sand	
Sand, fine-grained, pale mauve to buff brown, laminated, with muddy shaly layers up to 3 cm thick; thin bituminous layers	1.2
Clay, dark grey to black, with a 5 cm bituminous, yellow and mauve plastic clay at base	0.8 to 1.0
Sand, fine-grained, buff to yellowish brown, with pale brown marly clay beds up to 4 cm thick; thin ferruginous ribs	0.8
Sandstone, fine-grained, brown, concretionary, with some cross-bedding; irregularly channelled upper surface and an irregular base; strongly iron-stained in parts	0.3
Sand and sandrock, pale yellowish brown to lilac brown	0.5
Sand, coarse-grained, dark to medium brown, waterlogged; springs issue from base	0.3
Poole Formation	
PARKSTONE CLAY	
Clay, dark grey, bituminous, becoming finely laminated at depth, with sandy laminae up to 4 mm thick	2.0

Farther south, another section in the road cutting [SZ 1356 9641 to 1355 9637] shows 12.25 m of interlaminated, fine-grained, brown to lilac sand and grey clay in varying proportions, but with clay predominating; sand and sandstone intercalations are present, including a 0.55 m bed of brown sandstone 2.2 m above the base.

About 1 km south-west, an old pit [SZ 144 957] at the summit of St Catherine's Hill provides a good section in the Branksome Sand:

	Thickness m
Branksome Sand	
Sand, fine- to medium-grained, reddish brown, with some harder ferruginous beds	seen 1.0
Clay, pale lilac to bluish grey, with 10 cm of brown sandy clay at top	0.1 to 1.5
Sandstone and sand, fine-grained, brown, with some cross-bedding	up to 2.0
Sandstone, fine-grained, pale brown, thinly bedded	0.2
Sandstone, fine- to coarse-grained, hard massive, patchily red-stained	0.7
Sandstone, fine-grained, locally very coarse-grained, red-stained, cross-bedded, with ferruginous layers	1.2
Sand, fine- to coarse-grained, red-stained	1.5
Clay, white to pale grey; thickens eastwards within 20 m to include up to 1 m of brown sand, which itself thickens to the east	1.2 to 2.5

Some 360 m to the south, a large pit [SZ 1449 9523] reveals a succession, beneath 1 m of flinty gravel, that occurs lower in the formation than those given above:

	Thickness m
Branksome Sand	
Sand, fine- to medium-grained, pale to dark brown, iron-cemented in part, with thin (1 to 5 cm) white pipeclay seams	3.5
Sand, fine- to medium-grained, pale brown, locally cross-bedded, with harder clayey ferruginous beds	2.2
Sand, fine- to medium-grained, with bright red patchy staining; staining dies out below top 1.5 m	2.5
Sand, mainly fine-grained, but with very coarse-grained patches, brown, planar bedded; the base truncates the underlying beds where they are folded	1.0
Sand, mainly fine-grained, brown, with sandstone and ferruginous beds; bedding contorted and displaced by small-scale penecontemporaneous folding and faulting	3.2

BJW

In another old pit nearby [SZ 1453 9502], about 7 m of brown, fine- to medium-grained sand, with some coarse-grained patches, and thin ferruginous beds and locally contorted bedding are exposed. Traces of grey clay are seen at the top of the section. These strata lie below those of the previous pit. Another old pit hereabouts [SZ 1465 9494] shows the following section:

	Thickness m
Branksome Sand	
Sand, medium-grained, brown, with ferruginous cement at top	0.2
Clay, grey, passing down into clayey, very fine-grained, sand; probable palaeosol	1.0
Sand, medium-grained, brown to yellow, well-laminated at the top, with some cross-bedding; coarse-grained sand layers in the lower 1.5 m; erosional base	4.0
Sand, fine- to medium-grained, yellow to orange, with irregular silty clay areas; laminated silt at top	1.5

Two boreholes [SZ 1529 9657; 1548 9849] penetrated up to 4.8 m of laminated silty clay, interbedded with lignitic fine- to medium-grained sand, within the Branksome Sand beneath terrace deposits along the east side of the Avon valley (Clarke, 1981). ECF

Boreholes in the centre of Christchurch [around SZ 158 928] proved up to 8.5 m of river terrace deposits, above dominantly fine-grained, moderately well-sorted, dense sand. CRB

Beacon Hill

The Branksome Sand is exposed in the eastern part of the Beacon Hill pit [SY 9845 9520]. A typical section shows up to 10 m of fine-grained, thinly bedded sand, with beds of fine-grained sandy clay and clay. The clay beds, up to 3 cm thick, are commonly disrupted, and in places form clay-clast conglomerates with rounded pebbles up to 2 cm across. Some large-scale channels, 5 m deep, are present; small syndepositional faults with up to 10 cm displacement occur.

Brownsea Island

On Brownsea Island, the Branksome Sand is up to 15 m thick. Several sections show the junction with the Parkstone Clay. Those with the more extensive clay sequences are described in the account of the Parkstone Clay (pp. 53–54). A section [SZ 0135 8839] in the south-west of the island showed:

	Thickness m
Branksome Sand	
Sand, buff, medium- to very coarse-grained, with abundant clay clasts up to 0.3 by 0.1 m; erosive base	1.5
Sand, fine- to medium-grained, buff and orange, cross-bedded, with convolute bedding	4.0
Poole Formation	
PARKSTONE CLAY	
Clay, dark brown, very stiff	0.5

A section [SZ 0290 8743] on the south-east side of the island revealed:

	Thickness m
Branksome Sand	
Clay, brown, laminated, with wisps and fine laminae of very fine-grained sand; channelled base	0.4
Sand, brown, with impersistent clay-rich laminae	0 to 0.7
Sand, buff to orange, with clay laminae and layers up to 2 cm thick towards base; some disrupted bedding with some vertical laminae in part	0.7
Sand, fine-grained, buff, structureless	1.5
Sand, buff to orange, some thinly bedded, and cross-bedded; some channelling	0.5
Sand, buff to orange, structureless	3.5

Samples from the structureless sands are dominantly fine and medium grained, moderately sorted, and negatively to positively skewed.

A nearby section [SZ 0285 8741] exposed:

	Thickness m
Branksome Sand	
Clay, very silty, structureless, but with some 5 cm-thick laminae at base; thin, irregular, poorly sorted, medium- to coarse-grained sand beds, with clay clasts in some layers; some clay-clast conglomerates occupy small pods or channels; clay channelled into sand below	0 to 3.0

Another section hereabouts [SZ 0281 8742] showed:

	Thickness m
Branksome Sand	
Sand, medium- to very coarse-grained and pebbly, with a few clay clasts, comprising two fining-upward cycles; fine conglomerate at base of upper cycle 2 m from top	4.0
Clay, interlaminated with very fine-grained sand; each clay/sand unit of 1 to 2 cm thickness appears to fine upwards; lens of fine-grained sand 0 to 0.7 m thick, 0.5 m above base	3.5
Sand, medium- to fine-grained, poorly sorted, structureless in top metre, very thinly bedded in lower part	1.8
Clay, brown, with impersistent layers and laminae of very fine-grained sand	1.3
Sand, medium- to very coarse-grained; very coarse-grained sand bed at base	0.8
Sand, medium-grained, partly very coarse-grained, poorly sorted with some clay clasts; grit bed 0.15 m thick, 0.5 m from base	2.5

The lower clay unit passes eastwards over about 50 m of section into a more finely laminated clay, and then pinches out. This clay unit rests on the structureless clay seen in the previous section.

Some 20 m west of the last-described section, the lower clay unit is absent, probably having been cut out by the sand above, which contains large (20 cm) clay clasts at its base. A further 20 m to the west, the clay reappears, but the lamination is less pronounced. More than 0.5 m of dark brown lignitic clay, probably the lower clay of the last section, occurs about 1 m above beach level, but is cut out by a very coarse-grained gritty sand. A further 60 m west, the upper laminated clay and sand seen in the last section has passed into a sandy sequence, well laminated by grain size, in which finer grained layers are darker brown.

Yet another section [SZ 0265 8735] shows:

	Thickness m
8th River Terrace Deposits	
Gravel, orange-brown, with some cross-bedded sand units	up to 1.8
Branksome Sand	
Sand, buff, medium- to very coarse-grained, with some gritty layers; erosional base, with an undercut bank 1.5 m high trending N 300°	2.5
Sand, very fine-grained, interbanded with clay; some layers of fine-grained sand up to 15 cm thick; some signs of angular discordance	1 to 4.0
Sand, fine-grained, lenticular	0.5 to 0.8
Clay, dark brown, silty, lignitic, with wisps of very fine-grained sand; some lamination in the lower part	c.2.5
Sand, medium-grained, with scattered clay layers; cross-bedded in sets up to 1 m thick; some convolute bedding; coarsens to pebbly grit with rounded clay clasts at base; large scale cross-bedding also present. (This unit thickens westwards into a clay and lignite clast-rich conglomerate, with some carbonaceous laminae.)	5.0
Sand, buff, fine-grained, well-sorted	1.0

At the foot of South Shore Steps [SZ 0257 8738], a bed of fine-grained sand with clay layers and laminae occurs below the lignite and clay-clast conglomerate, and above the buff fine-grained sand.
ECF, CRB

River Stour to Poole Head

At Bearwood, the following section [SZ 0461 9602] showed the Branksome Sand resting in a channel cut into the sand beneath the Parkstone Clay:

	Thickness m
River Terrace Deposits	
Gravel, in pockets	0.2
Branksome Sand	
Sand, medium- to very coarse-grained, thinly bedded, with thin (up to 2 cm) beds of clay, and lenses of white silt 2 m long by 0.3 m thick	6.5
Sand, coarse- to very coarse-grained, planar- and cross-bedded, buff, with lumps and balls of white silty clay; sharp, slightly undulating base dipping 10° at N315°	2.5
Sand, medium- to coarse-grained, thinly bedded, becoming very coarse-grained westwards; some unit tops slightly clayey, others very coarse-grained; lower half cross-bedded; sharp planar junction dipping 12° at N255°	3.0
Poole Formation	
SAND BENEATH THE PARKSTONE CLAY	
Sand, fine-grained, locally clayey, white	1.5

Southwards, poorly exposed clay occurs at a higher level in the Branksome Sand. Mottled pale grey and orange silty clay occurs in the banks of a pond [SZ 0457 9553] on the golf course on Canford Heath. Some 450 m to the south [SZ 0458 9506], Reid (MS map, 1894) saw gravel overlying grey clay. The cryoturbated gravel/clay junction is exposed in a pit [SZ 0397 9516] on the south side of the Heath. Boreholes hereabouts proved up to 4.9 m of greyish brown mottled, fissured, silty clay with carbonaceous layers. Some 1.5 m of stiff, grey silty clay were proved in a borehole [SZ 0384 9525] to the north, beneath gravel.
CRB

A cutting [SZ 0475 9479] in the Ringwood Road showed, beneath river terrace deposits, some 5 m of yellow, medium- to coarse-grained sand with ferruginously cemented bands; the top 2 m are cross-bedded. The Branksome Sand occupies a north-east-trending channel cut into Parkstone Clay in a pit [SZ 0408 9480] at Alderney. There, 4 m of medium-grained, orange-brown sand rests with a sharply defined, ferruginously cemented base on Parkstone Clay; dewatering structures occur in the sand.

A poor section [SZ 0382 9427] in the Manning's Heath pit showed 0.5 m of ferruginous terrace gravel resting on 2 m of pale grey, orange-stained, roughly laminated, clayey, very fine- grained sand and silt. In a pit [SZ 0345 9362] to the south, the basal sand is well-to very well-sorted and fine- to medium- grained.

Lenticular developments of clay at about the same stratigraphic position as the clay on the golf course at Canford Heath (see above) occur at several localities. The most northerly is in a borehole [SZ 0415 9392] south of Alderney Hospital where, beneath 1.4 m of gravel, 2.4 m of grey silty clay rest on 0.5 m of yellow sandy clay, in turn resting on coarse- grained yellow sand.

South-west of Newton, a section [SZ 0330 9272] in an old pit exposed, beneath 1 to 2.5 m of river terrace deposits:

	Thickness m
Branksome Sand	
Sand, medium- to coarse-grained, well-sorted, thinly bedded, orange, cross-bedded	0.45
Pipe clay	0.05
Sand, medium- to fine-grained, becoming coarser-grained at depth, cross-bedded, with dewatering structures; brick-red staining near base; erosional base with ferruginously cemented layer	1.80
Sand, fine-grained, interlaminated with pale buff clay (1 to 2 mm layers)	0.08
Sand, well-bedded and laminated; some dewatering structures and cross-bedding; clay clasts in basal bed; erosional base	0.90
Sand, with disrupted clay layers	0.05
Silt, grey, clayey (pipeclay); laminated at top	0.40
Sand, fine-grained, rhythmically bedded in 0.05 to 0.08 m-thick units, with 1 to 2 mm seams of pipeclay; passes south into thicker-bedded, strongly cross-bedded units up to 0.15 m thick; foresets inclined 22° at N040° and 18° at N006°	over 1.50

Fine- to very fine-grained, very poorly sorted sands contain thin clay laminae in the old pit [SZ 0328 9262] south of the Old Wareham Road. A section [SZ 0370 9203] at Upper Parkstone, beneath 1.5 m of river terrace deposits, is as follows:

	Thickness m
Branksome Sand	
Sand, in fining-upward cycles around 0.6 m thick; the basal 0.1 to 0.4 m of the units consists of very coarse-grained sand which fines up into a medium-grained, clayey sand	1.2
Clay, grey, very silty	0 to 0.2
Sand, medium-grained, moderately sorted, reddish orange	0.8
Sand, very coarse-grained	0.05
Clay, grey, silty, red-stained	0 to 0.1
Sand, fine-grained, well-sorted, red-stained, with impersistent silty claybed about half way up; red staining dies out eastwards to give yellow sand. Some strongly convoluted laminae	4.0
Sand, fine-grained, well-sorted, yellow	0.3
No exposure	1.0
Sand, fine-grained, well-sorted, orange	2.0
Sand, very coarse-grained, and grit	0.1
Clay, grey, stiff	0.3
Sand, fine-grained, well-sorted, yellow	2.0

The basal beds of the Branksome Sand at Baden Powell Middle School [SZ 0366 9127] comprise 1.5 m of dark brown, lignitic, convoluted, clayey sand, resting on 0.3 m of buff, medium-grained, very well-sorted sand.

In an old pit [SZ 0390 9104] at Parkstone, the basal Branksome Sand consists of 1.3 m of buff, medium-grained, well-sorted sand, resting on Parkstone Clay. A section [SZ 0402 9094] 150 m south-east in this same pit showed:

	Thickness m
Branksome Sand	
Clay, silty, orange-buff and grey	0.1
Silt, clayey, orange to buff, becoming pale brown at depth	0.7

Sand, clayey, silty, lignitic, poorly laminated with chocolate-brown clay layers up to 0.3 m thick, some of which contain rootlets	over 1.2

A borehole [SZ 0451 9120] at Uplands School, Parkstone, encountered 10 m of coarse-grained sand with a 1.1 m bed of stiff, pale grey and brown silty clay in the middle. In the south of Parkstone, a section [SZ 0416 9029] exposed fine- to coarse-grained, cross-bedded sand and grit with thin (up to 1 cm) pipeclays.

A section [SZ 0431 8992] at the school at Lilliput showed 0.15 m of pale grey and orange silty clay, resting on 0.05 m of ferruginously cemented, medium-grained sand, which in turn overlies 2 m of yellow, medium-grained, cross-bedded sand. South of the school, up to 15 m of fine- to medium-grained sand has been proved in sections [around SZ 0430 8985] and boreholes.

Reid (MS map, 1894) noted 10 m of coarse-grained sand, very coarse-grained at the base, above the Parkstone Clay at Poole Head [SZ 0520 8855]. Silty clay, beneath gravel, occurs at the top of the cliffs [SZ 0550 8888] 450 m to the north-east.

Bournemouth cliffs

At the western end of the cliffs, between Poole Head and Branksome Dene Chine, mainly sand is exposed, varying from fine-to very coarse-grained and arranged in fining-upward cycles (A to G of Plint, 1983b) which commonly terminate in a bed of clay. Figure 16 shows the general sequence of these fining-upward cycles. The following section [SZ 0574 8913] low in Cycle A reveals, beneath 2 m of head, small fining-upward cycles forming part of the larger cycle A:

	Thickness m
Branksome Sand	
Sand, medium-grained	0.1
Clay, pinkish brown, very silty, becoming less silty at depth, with a ferruginous bed 0.2 m from top	0.6
Sand, very coarse-grained, yellow, well-cemented, fining upward to medium-grained; lateritic layer at top	0 to 0.4
Sand, fine- to medium-grained, yellowish brown	0.1
Sand, coarse- to very coarse-grained, poorly sorted, apparently within a channel	up to 0.2
Sand, yellowish brown, fine- to medium-grained, poorly sorted, becoming very coarse-grained downwards	over 0.3

Another section [SZ 0607 8932] in Cycle A at beach level exposed:

	Thickness m
Branksome Sand	
Clay, dark greyish brown, well-laminated and colour-banded; chaotic bedding from vertical to horizontal with large clay clasts; highly lignitic in places	1.0
Sand, very coarse-grained, subhorizontally bedded, with carbonaceous debris	over 0.6

This clay is not present in boreholes 150 m to the south-west.

A section [SZ 0755 9023] (Plate 5) at the top of Cycle C shows:

	Thickness m
Branksome Sand	
Clay, greyish brown to dark greyish brown, interlaminated and interbanded with thin sand layers which increase in thickness upwards; carbonaceous band at top.	c.6.0

This clay sequence passes westwards over 20–30 m into well-bedded, fine- to medium-grained sand in units of about 0.1 to 0.3 m thick, with only a few impersistent clay layers. The carbonaceous clay at the top also dies out westwards. There is some highly disturbed bedding.

The following section [SZ 0755 9025] at the top of the cliffs near Alum Chine, below river terrace deposits, typifies the fine-grained sediments at the top of Cycle F:

	Thickness m
Branksome Sand	
Sand, well-bedded in units 0.1 to 0.3 m thick	1.0
Clay, grey to buff, very silty, passing down into a clayey silt and finally into a very fine-grained sand; some ferruginous layers	1.3
Sand, fine-grained, well-bedded, grey to buff, with some clay content; interbanded with buff sandy clays with sharp bases in units of about 0.06 to 0.3 m., with some signs of coarsening upward; disturbed bedding and dewatering structures common; many lensoid sand units	2.2
Sand, medium- to very coarse-grained, grey, rather structureless	2.5

An excavation [SZ 0824 9058] for a cliff path east of Durley Chine exposed Boscombe Sand, beneath 5 m of made ground and drift, resting on beds comprising Cycle G:

	Thickness m
Boscombe Sand	
Sand, fine- to very fine-grained, well-sorted, grey	0 to 8.0
Sand, clayey, fine-grained, grey to buff	1.3
Branksome Sand	
Silt, clayey, buff and grey, blocky, fractured	1.0
Sand, very fine-grained, poorly sorted, buff	0.1
Sand, very fine-grained, clayey, brown, with an orange layer at base	0.4
Clay, silty, brown to greyish brown becoming dark brown towards the base	0.5
Silt, clayey, dark greyish brown, with dark brown roots	0.1
Sand, very fine-grained, clayey, roughly banded; banding picked out by rhythmic changes in clay content, which appear to be small coarsening-upward cycles varying from 0.05 to 0.01 m in thickness	1.3
Clay, roughly laminated with very fine-grained sand, which occurs in discontinuous lenticles and layers up to 1 cm thick; much carbonaceous debris	0.5
Clay, dark greyish brown, structureless, silty, with fine carbonaceous debris	0.5
Not exposed	1.5 to 2.0
Silt, clayey	c.4.0
Clay, grey and greyish brown, plastic (possible palaeosol)	0.3

Clay, silty to sandy and clayey sand, with some 0.2 m-thick ferruginous layers	2.0
Sand, medium- to coarse-grained, pale grey to buff, cross-bedded and with ferruginous layers; water seepage at base	over 3.0

The base of the section is 8 m above the promenade.

Farther east [SZ 0834 9057], beside the cliff lift, another section exposes clay at the top of the Branksome Sand:

	Thickness m
Boscombe Sand	
Sand, medium- to coarse-grained	over 0.2
Branksome Sand	
Clay, dark greyish brown, laminated, with very thin silty sand layers	6.0
Sandstone, fine-grained, ferruginous	0.15 to 0.5
Sand, medium- to very coarse-grained, orange and buff, cross-bedded, with ferruginously cemented layers	over 4.0

East of Bournemouth Pier, estuarine, bimodally cross-bedded sands and laminated clays (the Bournemouth Marine Beds of Gardner, 1879) of Cycle G are exposed. Near the pier, the following section [SZ 0917 9082] is exposed beneath head:

	Thickness m
Branksome Sand	
Sand, fine-grained, fairly well-sorted, cross-bedded, with many thin ferruginous bands	1.0
Clay, very silty, orange-stained, with an irregular top	6.0
Sand, fine- to medium-grained, with some clay masses	5.0
Sand, coarse-grained, yellowish orange, with lenticles of brown clay 0.6 m long by 0.08 m thick; some layers of very coarse-grained sand; ferruginously cemented band about 1 m above the base; trough cross-bedding common; sequence made up of minor fining-upward cycles on a scale of about 2 m	10.0

Some 360 m farther east [SZ 0953 9091], the following section occurs (probably all of the Branksome Sand lies in Cycle G):

	Thickness m
Boscombe Sand	
Sand, fine- to very fine-grained, yellow to orange, with many ferruginous layers	5.0
Branksome Sand	
Silt, clayey, brownish grey	1.3
Sand, fine-grained, silty, with clay beds and laminae	1.3
Sand, fine-grained, pale grey to buff, cross-bedded, with a ferruginously cemented layer at the top; some thin persistent clay laminae; coarse- to very coarse-grained sand layer about 2.7 m from the top; basal sand is also very coarse-grained	4.50
Clay, dark greyish brown, very lenticular	0.2
Sand, very fine-grained, interbedded with clay, in beds 0.05 to 0.80 m thick; the sand units thicken eastwards up to about 2 m	4.00

A borehole [SZ 0979 9093] at beach level proved 4.3 m of mottled greyish brown, silty, sandy clay, overlying 15 m of medium- to coarse-grained, in part gravelly, sand, with a few beds of sandy clay and traces of lignite. CRB, ECF

A section [SZ 1086 9113] in Cycle G in the lower part of the cliffs west of Boscombe is as follows:

	Thickness m
Branksome Sand	
Clay, olive-grey, homogeneous, with rootlets at the top; passing down over 0.6 m into a banded olive-grey clay and silty, very fine-grained, sand; each band of sand is a thin (1 cm) upward-coarsening unit	1.3
Sand, very fine-grained, pale brown, with lignitic debris and dark grey sulphide staining	0.8
Sand, very fine-grained, olive-grey to brown, banded with clay layers	1.0

A section [SZ 1105 9115] in Cycle G at the top of the formation, beneath the Boscombe Sand, occurs just west of Boscombe Pier:

	Thickness m
Branksome Sand	
Interlaminated and interbanded clayey silt and very fine-grained sand, with a few brown laminated clay layers; much fine-grained lignitic debris; some wedge-bedding	over 8.0
Sand, fine-grained, clayey, brown to orange, with disrupted streaks of clay	about 1.0
Clay, dark brown, lignitic, passing rapidly down into fine-grained, brown sand showing rough band ing, with bands of slightly more clay-rich carbonaceous sand; burrows, including *Ophiomorpha,* in the lower part	1.5

East of Boscombe Pier, the following section [SZ 1210 9126] of a 25 m-wide channel-fill in Cycle G is exposed: 0.1 m of brown clay, passing down into more than 2.5 m of brown lignitic sand, with a few thin brown clay layers. Another section [SZ 1250 9128], just to the east, shows:

	Thickness m
Branksome Sand	
Silt, clay and very fine-grained sand, roughly banded, brown to buff; clay beds impersistent; some bioturbation	1.3
Sand, fine-grained, lignitic, brown	1.0
Sand, very fine-grained, silty, yellow, with ferruginously cemented harder beds and some cream clay beds	0.5

About 1.22 km to the west, a borehole [SZ 1128 9111] entered the formation at 2.87 m below OD and proved the following:

	Thickness m
Branksome Sand	
Clay, firm to stiff, becoming stiff, fissured, dark grey, silty	2.9
Clay, silty, dark grey, laminated, with medium- to coarse-grained sand and partings of fine-grained sand and lignite	2.8
Sand, medium-grained, becoming coarse-grained with depth, grey; scattered thin very silty clay lenses	1.3

A section [SZ 1348 9127] in Cycle G 1.2 km to the east, near Southbourne, exposed:

	Thickness m
Branksome Sand	
Sand, fine-grained, clayey, with common brown clay laminae	over 3.0
Sand, fine-grained, yellowish orange, with some lateritic seams and some bands of ferruginous sandstone, particularly near the base	7.0
Sand, fine-grained, clayey, brown to orange, with disrupted clay laminae	1.0
Clay, dark brown, lignitic, passing down into fine-grained brown sand showing rough banding, with slightly more clay-rich carbonaceous sand. *?Ophiomorpha* burrowing in lower part	1.5

Further east, small exposures [SZ 1460 9112; 1533 9099] of laminated, dark brown clay and fine-grained sand occur at the top of the Branksome Sand. ECF

Hengistbury Head

The 1 m of laminated, dark brown, lignitic clay at the base of the cliff section [SZ 1660 9075] at Hengistbury (Hooker, 1975) is thought to correlate with the clay at the top of the Branksome Sand at Southbourne. In the BGS Hengistbury auger hole [SZ 1805 9072], 0.9 m of stiff, dark brown clay with disseminated patches and small (up to 1 cm) lumps of marcasite, was proved between 17.5 and 18.6 m below OD. A similar clay, up to 5 m thick, was proved in boreholes [SZ 1715 9235; 1697 9212; 1700 9168] on Stanpit Marsh and near Wick [SZ 1606 9170] (Freshney et al., 1984, p.21). A poor dinoflagellate microflora from the Stanpit boreholes includes species of *Apectodinium,* indicative of an inner neritic, possibly estuarine environment of deposition.

Gardner (1879) saw dark sandy clay with ironstone nodules in the bottom of The Run near Mudeford Ferry [c.SZ 1830 9155] and also farther north-east on the left bank of The Run. He thought that the nodules correlated with those in the Barton Clay at Hengistbury (see p.66), but Chandler (1963) stated that they were more like iron-cemented sands, and not like the Hengistbury nodules. Chandler obtained a limited flora from these beds. CRB

Wallisdown – Rossmore – Talbot Heath

An exposure [SZ 0552 9388] near Wallisdown showed 4 m of buff to yellow, fine-grained sand with a clay matrix in the lowest 1.5 m, underlying 1 m of river terrace gravel. A section [SZ 0637 9312] in an old clay pit south of Wallisdown proved, beneath 1 m of river terrace gravel:

	Thickness m
Branksome Sand	
Clay, silty, sandy, grey, with a blocky fracture, sand pods and lenses	2.0
Sandstone, coarse-grained, ferruginously cemented	0.2 to 0.6
Sand, medium- to very coarse-grained with some fine-grained, cross-bedded sand; thin (0.02 m) clay laminae and also some thicker seams (0.25 m)	2.0

A borehole at Rossmore [SZ 0531 9356] showed the following sequence, beneath 0.9 m of river terrace deposits:

	Thickness m	Depth m
Branksome Sand		
Sand, brown	1.83	2.74
Sand, clayey	0.92	3.66
Clay, sandy	0.61	4.27
Sand, pale grey	3.05	7.32
Sand, brown	1.52	8.84
Clay, white	0.61	9.45
Sand, pale grey	3.05	12.50
Poole Formation		
PARKSTONE CLAY	5.48	17.98
SAND BENEATH THE PARKSTONE CLAY	1.83	19.81

Further north, a borehole [SZ 0540 9497] penetrated over 3.6 m of stiff, fissured, greyish blue clay; this clay was extensively worked for bricks south of West Howe [SZ 052 961]. A borehole [SZ 0669 9322] at Bourne Bottom penetrated the following sequence beneath 2.25 m of sandy head:

	Thickness m	Depth m
Branksome Sand		
Sand, fine-, medium- and coarse-grained, dense, yellow, orange and brown, with sporadic fine gravel and clay lenses; abundant laminae of grey, orange and brown silty clay below 4.5 m, becoming common below 5.5 m	3.75	6.0
Clay, silty, stiff, greyish brown	0.90	6.9
Sand, fine- to medium-grained, silty, orange-brown, with laminae of grey silty clay	0.30	7.20

Ensbury Park – Winton – Meyrick Park

A borehole [SZ 0749 9491] at Ensbury Park proved:

	Thickness m	Depth m
River Terrace Deposits		
Gravel, compact, sandy	2.5	2.5
Branksome Sand		
Clay, pale yellowish grey, fine-grained, sandy, with orange clayey sand	0.3	2.8
Sand, yellowish brown, medium-grained	1.3	4.1
Silt, medium grey to greyish brown, slightly clayey and sandy	0.8	4.9
Clay, hard, dark brown, slightly silty	1.2	6.1

Another borehole [SZ 0823 9496] farther east penetrated a clay within Cycle G, beneath 3.3 m of river terrace deposits:

	Thickness m	Depth m
Branksome Sand		
Clay, hard, stiff, pale grey, silty, sandy, with bands of pale grey silty fine-grained sand	1.80	5.10
Sand, compact, yellowish brown, medium-grained, with bands of orange and pale grey silty clay; becomes slightly clayey with depth	2.05	7.15

ECF

Branksome – Westbourne

In a borehole near Branksome Station, the following sequence [SZ 0602 9180] was noted beneath 1.1 m of made ground and river terrace deposits:

	Thickness m
Branksome Sand	
Clay, silty, sandy, stiff, mottled grey, yellow, red and brown, with pockets of sand	4.0
Sand, silty, clayey, fine- to medium-grained, yellowish brown; becomes fine-, medium- and coarse-grained below 9 m, with fragments of ferruginously cemented sand; lenses of stiff white and brown, laminated, silty clay below 14 m; beds of yellowish brown, mottled silty clay between 23.1 and 23.8 m	24.0

Another borehole nearby [SZ 0627 9229] proved the base of the Branksome Sand at a depth of 30.5 m. In the Westbourne area, a further borehole [SZ 0764 9086] proved, beneath 4.27 m of river terrace deposits:

	Thickness m	Depth m
Branksome Sand		
Sand, coarse-grained, brown	1.83	6.10
Sand, coarse-grained, brown, with layers of clayey brown sand	3.20	9.30
Sand, medium- to coarse-grained, with layers of brown clayey sand	4.42	13.72
Lignite, firm, black	0.30	14.02
Clay, firm, brown, with traces of lignite	1.22	15.24

Central Bournemouth

Boreholes in western Bournemouth have encountered mainly fine- to medium-grained sand, but thin clay beds up to 0.9m thick occur locally [e.g. at SZ 0760 9137]. To the north-east, a borehole [SZ 0811 9151] proved, beneath 3.67 m of made ground and head: pale brown sand, 2.73 m; firm, grey clay, 0.92 m; grey sand, 3.35 m; firm, grey clay, 0.91 m. Another borehole [SZ 0850 9159] intersected strata high in the Branksome Sand, beneath 3.35 m of made ground and head, and proved:

	Thickness m	Depth m
Branksome Sand		
Silt, clayey, laminated, stiff, greyish brown, with layers of coarse-grained grey sand	5.03	8.38
Sand, coarse-grained, grey, with traces of lignite and thin layers of silt	5.64	14.02

A borehole [SZ 0878 9140] at Richmond Gardens car park penetrated:

	Thickness m	Depth m
Made ground	0.61	0.61
Boscombe Sand		
Sand, fine-grained, yellow	3.20	3.81
Branksome Sand		
Clay, silty, firm to stiff, brown and grey	2.59	6.40
Clay, very sandy	0.30	6.70
Sand, clayey, grey	1.53	8.23
Clay, firm, brown and grey	0.61	8.84
Clay, soft, grey	0.30	9.14
Sand, brown and grey, with clayey layers	3.66	12.80
Sand, with some ferruginous cementation	2.44	15.24

Nearer the centre of the town, boreholes [SZ 0878 9128; 0907 9134] penetrated several clay beds up to 1.15 m thick, as well as coarse-grained sand (Freshney et al., 1985, p.104). The latter borehole proved, beneath 3.35 m of made ground and head:

	Thickness m	Depth m
Branksome Sand		
Clay, brown, hard	1.15	4.50
Clay, sandy, brown	0.65	5.15
Sand, pale grey	3.25	8.40
Clay, sandy, pale grey	0.15	8.55
Sand, pale grey	1.35	9.90
Sand, yellow	0.40	10.30
Clay, sandy, pale brown	0.60	10.90
Clay, hard, grey	0.35	11.25
Sand, coarse-grained, yellow	1.00	12.25
Clay, sandy, pale coloured, with yellow sand	0.55	12.80
Sand, yellow	3.30	16.10

Another borehole [SZ 090 916] showed 6.25 m of clay with a band of claystone, below 1.22 m of Boscombe Sand, and overlying 3.05 m of fine-grained silty Branksome Sand. In the western part of the town, at West Hill, [SZ 0827 9106], the following succession was proved beneath 4.27 m of river terrace deposits: fine-grained, silty, brown sand with some lenses of grey sandy clay, 10.36 m; sandy, stiff to hard, pale grey clay, 1.22 m; on fine-grained, silty, brown sand 8.53 m. ECF, CRB

East of Bournemouth town centre, boreholes [e.g. SZ 097 916] show up to 2.4 m of laminated clay at the top of the Branksome Sand. One [SZ 0940 9152], penetrated a thicker clay sequence beneath 1.7 m of made ground and Boscombe Sand:

	Thickness m	Depth m
Branksome Sand		
Clay, stiff, silty, grey and yellow mottled	0.76	2.44
Sand, compact, silty, greyish yellow	0.61	3.05
Clay, silty, yellowish grey	0.50	3.55
Clay, dark brown	0.72	4.27
Clay, silty, sandy, brown	0.30	4.57
Clay, silty, yellow	0.46	5.03
Clay, silty, sandy, greyish yellow	0.61	5.64
Clay, soft, grey, with layers of yellow sand	0.76	6.40
Sand, silty, light yellow, with layers of grey clay	2.74	9.14

A well [SZ 098 917] in central Bournemouth penetrated a complete 69.8 m-thick succession of Branksome Sand beneath 8.4 m of drift and Boscombe Sand. The succession consists mainly of sand, but is capped by clay and includes several clay beds, one nearly 6 m thick and containing some lignite in the lower part (Freshney et al., 1985, p.182).

The clay at the top of Cycle G was extensively worked for brick clay near King's Park [SZ 1166 9270] and Queen's Park [SZ 1060 9328 and 1020 9340]. At Queen's Park, the following section [SZ 1064 9323] was recorded by Reid (MS, BGS): 3.1 m of laminated white clay, overlying 2.4 m of black loam, which in turn rested on 3.7 m of black loam and lignite. A borehole [SZ 1045 9200] hereabouts proved 1.85 m of stiff, dark brown silty clay, interlaminated with silt and sand partings, overlying 4.6 m of hard, brown, laminated, silty clay with partings and laminations of black and yellow sand. An exposure [SZ 1104 9317] below the clay of over 2 m of greyish brown, carbonaceous, medium- to coarse-grained sand with streaks of lignitic material occurs in a gully on the Queen's Park Golf Course.

A borehole [SZ 1071 9499] south of Mill Throop encountered stiff, greyish brown, laminated, silty clay, and fine- to medium-grained sand beneath river terrace deposits. ECF

SIX

Palaeogene: Barton Group

INTRODUCTION

The Barton Group consists of the Boscombe Sand, Barton Clay, Chama Sand and Becton Sand formations, in ascending order. At Hengistbury Head, the Barton Clay contains a sand member, the Warren Hill Sand (Freshney et al., 1984, p.46).

Hooker (1986) has shown that the Barton Clay, Chama Sand and Becton Sand consist of a series of coarsening-upward cyclothems representing marine transgressions and regressions. The Boscombe Sand, which was not studied by Hooker, represents an older cycle of transgression and regression. Hooker's (1986) ideal cyclothem consists, from the base up, of glauconitic clayey silt, glauconitic sandy silty clay, glauconitic silty clay, non- glauconitic clayey silty sand, and non-glauconitic sand.

BOSCOMBE SAND

The Boscombe Sand, which is 20 to 27 m thick, underlies much of Bournemouth and is well exposed along the cliffs. The beds form an outlier at Hengistbury Head and also occur in the cliffs at Highcliffe, where they dip eastwards beneath the Barton Clay. From the coast northwards to Ringwood, much of the outcrop is drift-covered.

The base of the Boscombe Sand is taken at a sharp lithological change, from laminated carbonaceous brown clay and associated palaeosol at the top of the Branksome Sand, to fine- to medium-grained, well-sorted sand. Where the highest bed of the Branksome Sand is a sand, it is usually silty and clayey, and the boundary is then an erosion surface, marked by contrasting sedimentary and grain-size features in the sands; the boundary was regarded as a marine transgression surface (T4) by Plint (1983b).

The Boscombe Sand comprises mainly fine- to medium-grained sand, consisting dominantly of quartz grains, but at some levels flint grains are common. The mean grain size varies between medium (1.0ø) and very fine (3.5ø), with an average of fine (2.2ø). The sorting ranges from very well sorted (0.18ø) to poorly sorted (1.7ø), with an average of well sorted (0.42ø), and the skewness between + 0.7 and − 0.5 with an average of + 0.1. Some samples, notably those from the top of the Boscombe Sand at Hengistbury Head, have a sorting average of 0.3ø and a skewness average of − 0.4. Scatter plots of mean size/sorting, and sorting/skewness are shown on Figures 13 E and F.

The Boscombe Sand exhibits two main bedding forms: planar laminated beds with a few planar cross-bedded units containing flint pebble and cobble beds in the lower part, which are associated with marine sands; and bi-directional cross-bedded strata in the upper part, which characterise estuarine sands.

In the Christchurch Borehole, the Boscombe Sand has been divided into two units, each showing different characteristics in the gamma-ray log. The lower 9 m show an upward increase in mean grain size by the loss of the clay content. The upper 10.3 m are more clayey, particularly in the topmost few metres. In the adjacent cliffs and at Hengistbury Head, the uppermost part of the Boscombe Sand contains upward-coarsening, carbonaceous, silty, bioturbated sands, and remobilised muddy sands (slurry beds) containing clasts of bituminous sand (Figure 17).

The pebble and cobble beds, which are composed of rounded to well-rounded flint pebbles, are well displayed in the cliffs between Boscombe and Southbourne, and inland at Matcham's Park [around SU 125 020]. Good sections in the pebble and cobble beds and overlying fine-grained sand of the Boscombe Sand occur on Hengistbury Head where, locally, bituminous sands up to 2.5 m thick are present. A section through a channel containing these bituminous sands is shown in Figure 17. The origin of this bituminous sand, its subsequent disruption by liquefaction, possibly as a result of seismic shocks, and its incorporation in the 'slurry beds' is described by Plint (1983c).

At Friars Cliff [SZ 197 928], Plint (1983b, fig.9) described three upward-coarsening, estuary mouth-bar sequences, each 5 to 6 m thick. Each unit consists of an upward gradation from intensely bioturbated, dark brown, sandy and silty clay, rich in plant debris, through silty sand, into fine-grained, clean, faintly parallel-laminated sand. This last is overlain by cross-bedded, fine- to medium-grained sand containing layers of mud pellets. A sand, 1.4 m thick, at the top of the Boscombe Sand in this area [SZ 1980 9287] appears to have been destructured through quicksand action. Buff sand, up to 8 m thick, with large- scale dewatering and ball-and-pillow structures occurs farther east [SZ 199 928] (Plate 6). Most of the balls are between 0.5 and 1 m across, but some are up to 4 m in diameter. The dewatering structures penetrate 4 m of beds.

At the western end of Hengistbury Head, an impersistent layer, up to 1.25 m thick, of well-sorted, fine-grained sand with flint cobble beds at the top and bottom occurs at the top of the Boscombe Sand. Thin (1 to 100 mm) pale grey to cream, silty clay seams (pipeclay) are fairly common throughout the sequence.

The Boscombe Sand has yielded few body fossils. The topmost beds at Hengistbury Head and Southbourne contain dinoflagellate cysts indicative of the *intricata* Assemblage Zone (Costa et al., 1976).

Chandler (1963, p.18) obtained more than fifty species of plants, together with the brackish or marine bivalve *?Meretrix,* and the foraminifer *Nummulites* sp. from the Boscombe Sand at Friars Cliff. Burton (in Chandler, 1963) recorded unspecified molluscs from the same beds.

Faunal and floral evidence from outside the district suggests that the Boscombe Sand is contemporaneous with marine clays in the New Forest area (the Huntingbridge Formation of Curry et al., 1978). ECF, CRB

BARTON CLAY

The formation crops out in the east from the coast at Highcliffe, northwards to Crow [SZ 170 037], east of Ringwood; most of the outcrop is obscured by river terrace deposits. There is an outlier at Hengistbury Head, and an inlier near Burley in the north-east.

The formation, 26 to 60 m thick, consists of mainly yellow-weathering, greenish grey to olive-grey, commonly glauconitic clay, with a variable content of both disseminated and bedded, very fine-grained sand, particularly at the western end of Hengistbury Head (p.72). Ironstone nodules are particularly common at Hengistbury Head where they occur at four levels (Figure 18). The nodules were quarried and also collected from the foreshore for iron-making between 1847 and 1865 (Tylor, 1850; West, 1886). The Barton Clay is usually strongly bioturbated, with little sign of lamination. Where unweathered, the clays are commonly shelly; the upper part of the formation is particularly rich in bivalves.

Above the glauconitic clays in the cliff section at Warren Hill, Hengistbury, there are about 10 m of very fine-grained, buff and yellow, unfossiliferous, cross-bedded sands overlain by river terrace deposits (Plate 8). The sands were first noted by Lyell (1827), but not named until 1879 when Gardner referred them to the Highcliff [sic] Sands on the basis of their supposed correlation with sands (the Boscombe Sand) that crop out under the Barton Clay at Highcliffe [SZ 200 929]. A new name, *Warren Hill Sand,* was proposed after the type section at Warren Hill [SZ 1700 9050], Hengistbury, by Freshney et al. (1984, p.46). Although thin beds of lithologically similar sand occur in the Barton Clay beneath the Warren Hill Sand, the base of the latter appears to be sharp; the junction is only exposed in the upper part of the cliff and is not easily accessible.

Various definitions of the base of the Barton Clay have been proposed. Prestwich (1849) took it at the base of the pebble bed beneath a glauconitic sandy clay at Highcliffe Castle [SZ 203 931]. Keeping (1887), followed by Curry et al. (1978), adopted the incoming of the foraminifer [*Nummulites prestwichianus*] as the base of the Barton Clay, at an horizon about 3 m above that chosen by Prestwich, and within an apparently uniform glauconitic sandy clay. At Highcliffe, Prestwich's basal bed consists of well-rounded black flint pebbles, set in a sandy glauconitic clay. At Hengistbury, a similar clay locally rests on a cobble gravel at the top of the Boscombe Sand (Freshney et al., 1984, fig.7; Plate 7). In places hereabouts, the cobbles have been incorporated into the basal bed of the Barton Clay; thus it resembles the pebble bed at Highcliffe. Elsewhere at Hengistbury, the cobble bed is only locally well developed and the Barton Clay commonly rests directly on clean, very well- sorted, very fine-grained sand, up to 1.25 m thick, with a lower cobble bed locally well developed at its base. Plint (1983c, fig. 2; 1983b, fig. 6) regarded the base of the cobble bed beneath the fine-grained sand (the base of his T5 transgression) as the base of the Barton Clay.

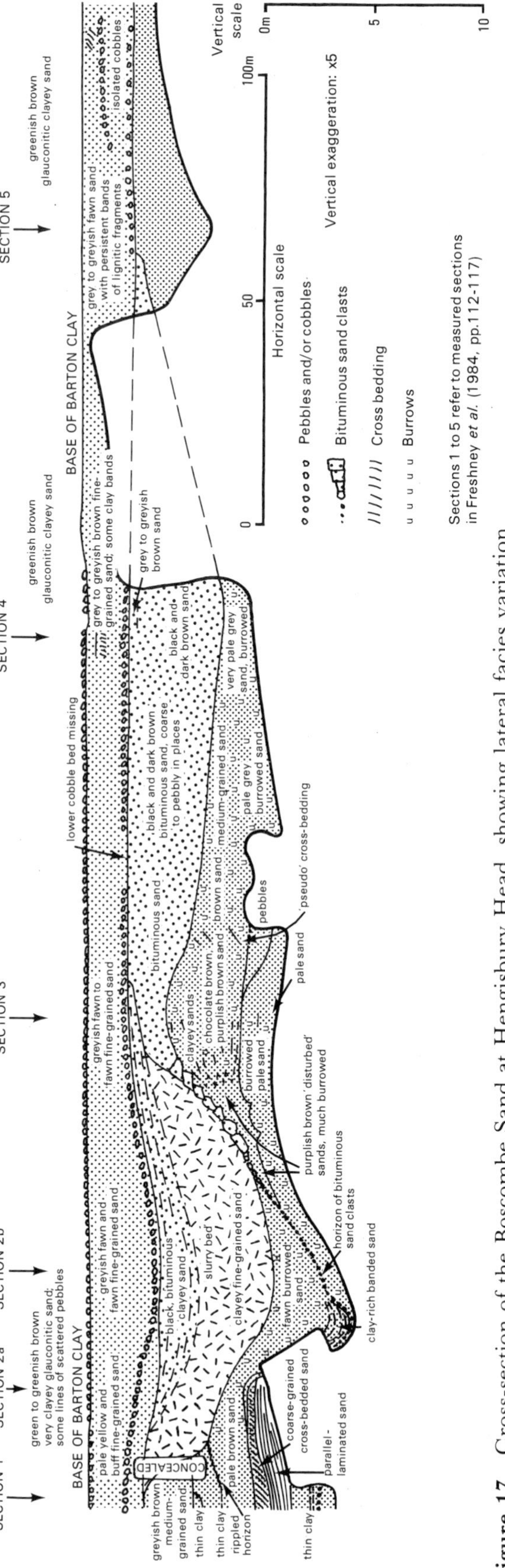

Figure 17 Cross-section of the Boscombe Sand at Hengisbury Head, showing lateral facies variation.

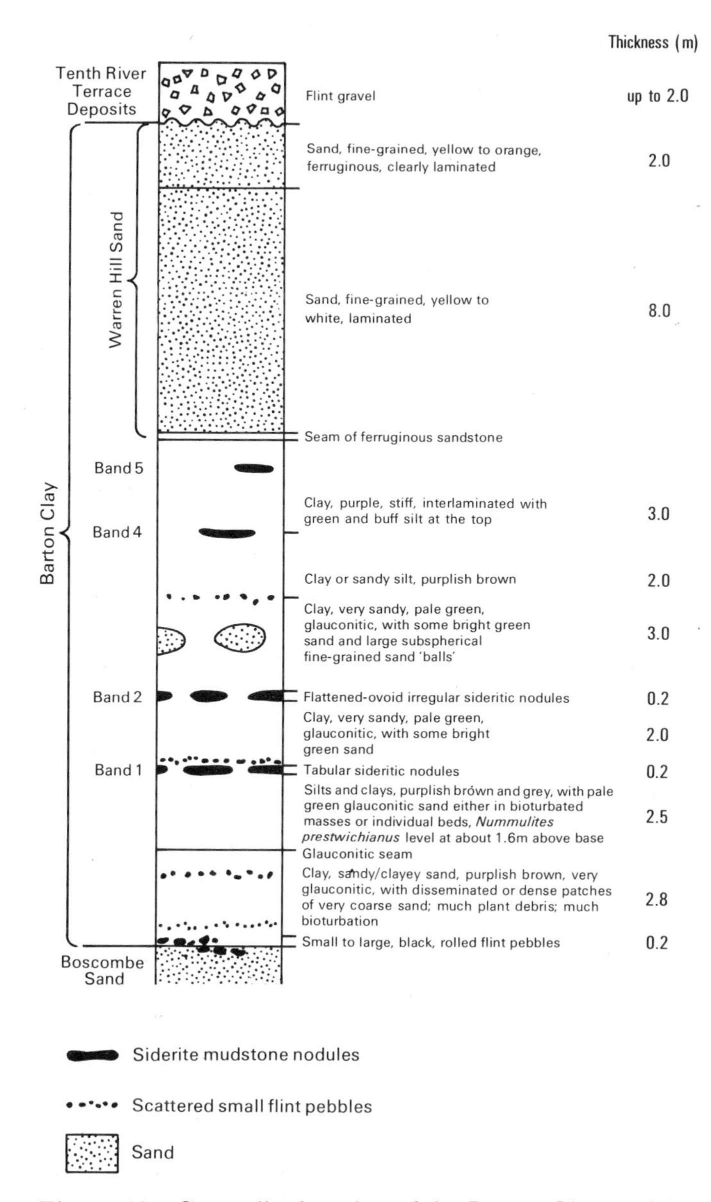

Figure 18 Generalised section of the Barton Clay at the western end of Hengistbury Head (after Hooker, 1975).

For mapping purposes, a boundary taken either at the level of the incoming of *N. prestwichianus* (following Keeping, 1887), or at the base of the lower cobble bed (following Plint, 1983b; c) is impractical. The present authors have therefore followed Prestwich's (1849) definition, because it is based on a persistent lithological change that can be traced inland.

The problem of whether the glauconitic sandy clays at Hengistbury (the 'Hengistbury Beds' of Gardner, 1882) are at the same stratigraphical level as similar strata, the Barton Clay, exposed at Highcliffe [SZ 200 929], has been the subject of much discussion since the time of Lyell (1827). The controversy was summarised by Hooker (1975). The currently accepted interpretation, first advocated by Prestwich (1849), is that both clays are at the same stratigraphical level. This was supported palaeontologically by Curry (1942) and by Costa et al. (1976), and by the heavy mineral analyses of Blondeau and Pomerol (1969). The present survey indicates that the west-north-west-trending Christchurch Fault is responsible for repeating the Barton Clay sequence at Hengistbury Head.

Burton (1933) divided the Barton Clay of the type area at Barton on Sea [SZ 235 929] into a number of faunal and lithological divisions, numbered in ascending sequence A1 to A3, and B to F respectively (Figure 8). For the most part, these are not mappable units, although unit A3, a grey clay with beds of fine-grained grey sand, visible in the cliffs at Barton on Sea, is probably the lateral equivalent of the Warren Hill Sand at Hengistbury.

Blondeau and Pomerol (1969) studied the heavy minerals from the Barton Clay at Hengistbury. In the higher beds, kyanite and garnet are more abundant than staurolite, and become increasingly abundant upwards, where they are associated with a rich assemblage of epidote, anatase, brookite, hypersthene and titaniferous corundum. The presence of epidote suggested to them a correlation with part of the lower Barton Clay at Barton on Sea. The clay minerals of Hengistbury include kaolinite (50 to 80 per cent), with subordinate illite, although in one sample the proportions are reversed.

The clays at Hengistbury are only sparsely fossiliferous. Nevertheless, the fauna is diverse and includes plants, foraminifera, molluscs, crustaceans, echinoids and fish (Chandler, 1960; Chapman, 1913; Curry, 1942; Hooker, 1975; Reed, 1913; Stinton, 1975; 1977). The type section at Barton on Sea [SZ 235 929] contains an abundant marine fauna, dominated by gastropods and bivalves, with corals, serpulids, scaphopods, echinoids and fish vertebrae. The fauna is indicative of shallow marine conditions of normal salinity.

Work on the dinoflagellate cysts has shown that the beds up to the level of the *N. prestwichianus* nummulite horizon belong to the *intricata* Assemblage Zone of Bujak et al. (1980), and that the beds above belong to the succeeding *draco* Zone (Costa et al., 1976). ECF, CRB

CHAMA SAND

The Chama Sand is only patchily exposed below the drift cover in the south-eastern part of the district, but the outcrop is more extensive near Burley [SU 215 032] in the north-east. It is about 8 m thick at Barton on Sea, 6 m thick near Bransgore and 10 m thick around Burley. The formation consists of greenish grey to grey, slightly glauconitic, clayey, silty, very fine-grained sand, silt and sandy clay. Where unweathered, it is commonly shelly. The formation is highly bioturbated; in places, there are near-vertical burrows, possibly of *Ophiomorpha*. The Chama Sand forms a transitional unit between the Barton Clay and the overlying Becton Sand. Its base is marked by the incoming of slightly glauconitic, clayey, silty, very fine-grained sand, which in the field coincides with a concave topographical feature and a spring line.

The Chama Sand has a mean grain-size ranging from very fine sand (3.3ø) to coarse silt (5ø); it shows moderate to very poor sorting (sorting index = 0.5 to 2.31ø), and grain-size distributions that are strongly positively skewed (skewness coefficient = 0.5 to 0.8). The heavy minerals are

Plate 6 Barton Clay overlying Boscombe Sand with pebble bed at junction, Friar's Cliff. 'Ball and pillow' structures resulting from dewatering occur at the top of the Boscombe Sand.

composed dominantly of Morton's (1982) Association A, i.e. a garnet-epidote assemblage derived from a northern metamorphic basement. ECF

BECTON SAND

The Becton Sand crops out extensively in the Burley area, and southwards towards Bransgore where it forms well-drained land. South-east of Bransgore, it is largely obscured by terrace gravels. The thickness of the Becton Sand near Burley is about 20 m, at Bransgore about 7 m, and at Barton on Sea 24 m. The best exposures lie just outside the district in the cliffs at Barton on Sea where the formation is divided into three members. The middle, Becton Bunny Member, comprising greyish brown shelly clay, is impersistent and has only been seen in one inland exposure. The beds above and below consist of fine-grained sand and cannot be differentiated in the absence of the Becton Bunny Member.

The base of the Becton Sand is marked by the absence of clay and silt, which are characteristic of the Chama Sand. In surface exposures, the Becton Sand consists of yellow to pale grey, well-sorted, fine- to very fine-grained sand composed dominantly of angular to subrounded quartz grains. At most localities, the sand appears structureless, although cross-bedding is present locally; the lower part is bioturbated and shelly. Callianassid burrows and rootlet horizons have been recorded outside this district.

The mean grain size of the sands varies between fine (2.4ø) and very fine (3.7ø) (Figures 13G and H). In a few localities near the top of the formation, some medium-grained sand occurs. The sorting varies between very well sorted and moderately well sorted (sorting index = 0.18ø–0.90ø). The Becton Sand usually shows low positive to negative skewness values. The heavy minerals (Morton, 1982) are dominated by northerly derived material characterised by garnet and epidote. ECF

Conditions of deposition

The onset of Boscombe Sand sedimentation represents a marine transgression, which Plint (1983b) equated with his T4 Transgression. The formation was deposited in conditions ranging from marine beach and shoreface in the lower

Plate 7 Junction of Barton Clay and Boscombe Sand, Highcliffe. Glauconitic sandy clay at the top of the section has a pebble bed of well-rounded flints at the base.

part, to estuarine channels with a sediment flow from the west and west-south-west in the upper part.

The lower unit of the Boscombe Sand in the Christchurch Borehole probably represents a marine transgression followed by shoreface progradation, whilst the upper unit was deposited in a tidal estuary. The 'slurry beds' near the top of the formation were regarded by Plint (1983c) as estuary mouth-bar sands and estuary channel plugs respectively. The cobble beds at the top of the formation probably indicate a return to shoreface conditions, the cobbles having been derived from a storm beach. Plint (1983b) regarded the lower cobble bed as marking his T5 Trangression.

A third marine transgression within the Barton Group is marked by the glauconitic clays of the Barton Clay. The mean grain size, poor sorting, strong positive skewness of the grain-size distribution, bioturbation and the shelly fauna of the clayey sands indicates deposition on a marine shelf below normal wave base. Hooker's (1986) cyclothems represent minor transgressions and regressions; the regressions resulted in the deposition of shoreline facies. The most important of these is the Becton Sand, at the top of the group, in which bioturbation, including callianassid burrows, rootlet horizons at the top, and its grain size and bedding characters indicate deposition in the upper shoreface to upper beach zones. Some aeolian sand also may be present. A postdepositional palaeosol is commonly developed at the top of the Becton Sand, immediately beneath the Headon Beds. ECF

Details

BOSCOMBE SAND

Bournemouth area

West of Durley Chine, a temporary section [SZ 0824 9058] exposed up to 8 m of fine-grained, well-sorted sand overlying Branksome Sand. West of the pier, a cliff section [SZ 0953 9091] exposes over 5 m of fine- to very fine-grained sand with ferruginous layers. Local lenticular beds of well-rounded flint gravel occur high in the cliffs [e.g. SZ 0986 9100].

Inland exposures are rare and most information comes from boreholes (Freshney et al., 1984). East of Bournemouth town centre, a borehole [SZ 0953 9101] showed the following section beneath blown sand and river terrace deposits:

Plate 8 Warren Hill Sand Member, Hengistbury Head. A fine-grained orange to buff sand rests on grey sandy clay of the Barton Clay. The ironstone nodules in the Barton Clay were once worked as iron ore.

	Thickness m	*Depth* m
Boscombe Sand		
Sand, fine-grained, compact, yellow	5.34	12.04
Sand, fine-grained, silty, pale grey, with thin silt layers	1.68	13.72
Sand, compact, pale grey, silty	3.73	17.45
Sand, fine-grained, pale brown, with layers of yellow silt	0.84	18.29
Clay, grey, mottled, silty, sandy	0.61	18.90
Sand, fine-grained, silty, clayey, yellow	0.30	19.20
Silt, compact, grey, sandy	4.88	24.08
Branksome Sand		
Silt with some clay and ferruginous cemented sand	0.30	24.38

Exposures of fine-grained sand, overlying coarse-grained Branksome Sand, occur on the valley sides [around SZ 084 922] in Meyrick Park. West of Bournemouth town centre, a borehole [SZ 09819 9073] showed the following succession beneath 4.19 m of river terrace deposits: fine- to medium-grained, dense, brown sand, 11.05 m; on medium-grained, reddish brown, dense sand, 1.37 m; on Branksome Sand.

North-east of Boscombe Pier, a section [SZ 1133 9125] exposes 4.5 m of buff to grey, fine-grained sand with many 2 to 5 mm-thick clay-rich layers. Exposures on the cliff top [around SZ 120 913 and 1319 9129] show yellow, fine-grained sand with a well-developed flint cobble bed, the cobbles being up to 10 cm in diameter. Another section [SZ 1370 9127] 50 m to the east exposes:

	Thickness m
Boscombe Sand	
Cobble bed of flints	1.0
Sand, fine-grained, yellow to orange, with some ferruginous bands	1.5
Sand, fine- to very fine-grained, well-sorted, with some thin ferruginous bands and some thin clay layers	3.0

Thin beds of bleached and very well-rounded pebbles in a matrix of fine-grained sand, overlain by very clayey fine-grained sand, occur farther east on the cliff top [SZ 1467 9115]. ECF

Hengistbury Head

Good exposures of Boscombe Sand occur in the western part of Hengistbury Head [SZ 178 904]. There, it consists dominantly of buff, brown or greyish brown, fine- to medium-grained, laminated, or locally cross-bedded, sand; burrowing is common. Stringers of pebbles occur in the lowest strata exposed; thicker concentrations of pebbles and cobbles occur at the top of the sand and these persist laterally for at least 350 m eastwards in the cliff sections. Towards the top of the formation, there is a bituminous-stained, weakly cemented sandstone, up to 2.5 m thick (Figure 17). At the western end of the section, fragments of this sandstone lie in the base of a channel filled with muddy sand. The origin of this bituminous bed and its subsequent disruption is described by Plint (1983a). The sands in the channel, and also others outside the channel below the lowest cobble bed, are well sorted, dominantly positively skewed and fine grained. In contrast, the sands between the cobble beds are very well sorted, negatively skewed and very fine grained. CRB

Highcliffe–North Ripley area

In the Christchurch Borehole [SZ 2002 9301], glauconitic, carbonaceous, bioturbated, brownish grey, fine- to medium-grained, slightly clayey sand was recovered in the cored section between 16.20 m and 20.45 m. The uncored section below (20.45 to 30.23 m) appears, from the gamma log, to be a cleaner sand than the beds above and, by comparison with exposures west of Christchurch Harbour, is probably a well-sorted, fine-grained sand.

At Friars Cliff [SZ 197 928] west of the borehole, where the formation is up to 18 m thick, up to 11 m of fine-, medium- and coarse-grained, buff and brown sand are exposed. A typical section [SZ 1980 9287] comprises:

	Thickness m
River Terrace Deposits	
Gravel, sandy	0.7
Barton Clay	
Clay, very sandy, fine-grained, glauconitic	1.5
Pebble bed of well-rounded black flints within a glauconitic sandy clay	0.1
Boscombe Sand	
Sand, fine-grained, structureless, buff	1.4
Sand, fine-grained, thinly bedded, buff	8.0
Sand, clayey, carbonaceous, dark greyish brown	0.4
Pebble bed with buff, rotted, patinated flints	0.02
Sand, fine-grained, buff	0.3

The distinctive 0.40 m-thick carbonaceous sand thins eastwards to about 0.20 m some 60 m east of the section; a further 60 m to the east, this bed descends to beach level [SZ 1992 9290]. The structureless sand at the top of the Boscombe Sand appears to have lost its depositional features through quicksand action. Some 60 m to the east [SZ 1986 9288], this uppermost bed consists of 8 m of fine-grained, buff sand with ball-and-pillow structures. Most of the balls are between 0.5 and 1 m in diameter, but some are up to 4 m. Large-scale dewatering structures are present through 4 m of beds.

The Friar's Cliff section is difficult to relate in detail to Plint's (1983b, fig. 9) section, in which he described three coarsening-upward sequences varying in thickness from 3.5 to 5.5 m. Most of the section probably corresponds to his uppermost sequence, although the pebble bed towards the base of the section, which Plint placed within the lower part of his upper sequence, is more likely to mark the top of the underlying coarsening-upward cycle. Each of Plint's units consists of an upward gradation from intensely bioturbated dark brown, sandy and silty clay, rich in plant debris, through silty sand into fine-grained, clean, horizontally laminated sand. This sand is overlain by cross-bedded, fine- to medium-grained sand containing layers of mud pellets. However, in the Christchurch Borehole 200 m to the east, the gamma-ray log shows evidence for a basal 9 m-thick coarsening-upward sequence, succeeded by a 2.5 m-thick coarsening-upward sequence, which in turn is overlain by 5 m of sand and clayey sand with no evident cyclic gradation, capped by a 3 m-thick fining-upward sequence. CRB

A power auger hole [SZ 1967 9517] at Burton Common penetrated the base of the Barton Clay at a depth of about 8 m and then proved 16 m of brown sand and fine-grained sandy clay. An old gravel pit [SZ 1830 9702] at Bransgore exposed over 2 m of ferruginous orange-brown, medium-grained, sand beneath river terrace deposits.

Between North Ripley [SU 170 000] and the edge of the district, the Boscombe Sand outcrop is largely obscured by river terrace deposits, but up to 3 m of carbonaceous silty, fine- to medium-grained, locally clayey sand was proved in boreholes [SZ 1662 9931 and 1745 9664] beneath up to 6.9 m of river terrace deposits.

ECF

St Ives–Matcham area

An exposure [SU 1255 0393] at David's Hill reveals beds close to the base of the formation:

	Thickness m
Boscombe Sand	
Sand, fine-grained, pale brownish grey, locally cross-bedded, with thin iron-rich layers	3.15
Clay, greyish brown, silty	0.25
Sand, fine-grained, greyish brown, with thin clayey beds	0.15
Sand, fine-grained, pale brown, cross-bedded, with thin iron-cemented layers	4.50

At the side of a track [SU 1290 0102] in Matcham's Park, the following exposure occurs: fine-grained, buff, structureless sand, 1.25 m; fine-grained, pale brown, thinly bedded sand, 1.05 m; on pale brown, cross-bedded sand, 0.95 m. About 300 m south-east, small exposures in a track surface [SZ 1298 0082] reveal some 8 m of pale brown, carbonaceous, silty, bioturbated sands, which coarsen upwards in units 0.1 to 0.4 m thick. Thin impersistent silty clay beds less than 1 cm thick also occur.

Some 200 m north of Foxbury Hill [SU 1221 0070] an exposure shows:

	Thickness m
Boscombe Sand	
Pebble bed, bleached, with well-rounded, ovoid flint pebbles from 2 to 6 cm in diameter, mostly about 4 cm; matrix of brown fine-grained sand	0 to 0.3
Sand, brown and pale brownish grey, fine-grained, with thin ferruginously cemented beds	0.3 to 0.6
Pebble bed, as above, clast-supported, with poorly developed imbrication suggesting transport from the west	0 to 0.40
Sand, mainly fine-grained, with some medium-grained beds up to 0.1 m thick; sporadic thin ferruginous beds	9.5

BJW

Barton Clay

Sandford – Highcliffe

A borehole [SU 1696 0104] near Sandford proved 2.3 m of silty, mottled orange and grey clay which is greyish blue at depth, under 1.2 m of clayey gravel. In Ripley Wood, another borehole [SU 1629 0023] showed 1.6 m of sandy, mottled green and dark grey clay, and clayey sand, below 5.4 m of terrace deposits.

A dutch auger hole [SZ 2183 0359] near Burley Lawn revealed 2 m of yellow, passing down into greenish grey clay, overlying 0.3 m of greenish grey, silty, shelly clay. Another auger hole [SZ 2384 0327] proved 0.6 m of clayey gravel, resting on over 1 m of brown-weathering, greenish grey, shelly clay.

A stream section [SZ 1827 9727] at Bransgore showed 0.7 m of yellowish orange and brown, extremely sandy clay and clayey sand with a basal bed, 0.05 m thick, of well-rounded flint pebbles overlying Boscombe Sand. At Bransgore [SZ 1838 9776], Reid (MS, BGS) recorded 10 m of clay with ironstone nodules, overlying 1.1 m of clay. A stream section south of this [SZ 1824 9725] showed slightly glauconitic, extremely sandy clay with rounded flints at the base, overlying Boscombe Sand.

A borehole [SZ 2084 9602] at Hinton Admiral House penetrated the entire Barton Clay sequence, which is 53.64 m thick. The formation consists of 14.02 m of blue clay with traces of shells, overlying 39.62 m of sandy, green, brown and blue clay.

The BGS Burton Common No. 2 Borehole [SZ 1958 9495] proved 3.17 m of extremely sandy, glauconitic, greenish grey clay beneath river terrace deposits and above Boscombe Sand. In nearby Burton Common No.1 Borehole [SZ 1967 9517], about 8 m of glauconitic, extremely sandy clay on Boscombe Sand were proved. ECF

Glauconitic sandy clay occurs in the banks of the sunken path [SZ 1967 9305] near the Golf Course, and in the road banks [SZ 1952 9337 to 1959 9340] to the north-west.

Barton Clay was formerly exposed along the cliffs from Highcliffe eastwards, but the section is now mostly obscured by slips or cliff protection works. The top of the clay is commonly cryoturbated into the overlying gravel. Scattered sections that have been intermittently exposed in recent years by landslipping, and fossils collected from them, have been recorded by Clasby (1971; 1972), Daniels (1970) and Hooker (1975). A maximum of 6 m of glauconitic sandy clay with scattered cementstones, and a basal pebble bed, are exposed beneath 2 m of river terrace gravel at Highcliffe Castle [SZ 2080 9314]. Farther east, a section [SZ 2156 9315] exposed 2 m of river terrace sand and gravel above 1 m of glauconitic silty clay with a brown weathered top, resting on a 0.3 m-thick shelly sandstone, which in turn overlies 3 m of glauconitic, fine-grained sandy clay. CRB

Hengistbury Head

The Barton Clay of the cliff sections at Hengistbury has been described by many authors (Lyell, 1827; Prestwich, 1849; Gardner, 1879; Reed, 1913; White, 1917; Hooker, 1975; Curry, 1976).

The bulk of the Barton Clay at Hengistbury is a glauconitic, very clayey, fine-grained sand, or very sandy clay. A generalised section for the Barton Clay at the western end of Hengistbury Head (based on Hooker, 1975) is given in Figure 18. Beds of fine-grained sand are particularly common in the west, but most beds thin eastwards and pass into sandy clay. One locality [SZ 1782 9040] at the top of the cliff in the east, however, revealed 0.7 m of fine-grained, greyish buff sand above glauconitic sandy clay. Farther east, a section [SZ 1804 9052] exposed a bed of fine-grained sand, 0.5 to 1 m thick, within glauconitic sandy clay and clayey sand. Small (up to 1 cm across) fragments of lignite are common at this locality, together with moulds of bivalves.

The ironstone nodules, a characteristic feature of the Barton Clay at Hengistbury, are up to 1 m in diameter and occur as impersistent layers at four levels; numbers 1, 2 and 4 persist throughout the section (Figure 18). Balls of well-sorted, very fine-grained sand, up to 1 m across, occur in the middle part of the sequence.

Poor exposures of glauconitic sandy clay with ironstone nodules occur in the sides of the old ironstone quarry [SZ 1741 9054 to 1765 9065] at Hengistbury. Farther west, a section [SZ 1681 9080] in the basal beds revealed 1 m of medium-grained, clayey, glauconitic sand resting on a ferruginously cemented pebble bed, 0.1 m thick, at the top of the Boscombe Sand. A second section [SZ 1685 9081] exposed 0.6 m of glauconitic sandy clay with a layer of small (15 mm-diameter), well-rounded, black flint pebbles 0.4 m above the base, resting on a pebble bed with a coarse-grained, ferruginously cemented sand matrix, 0.1 to 0.2 m thick, at the top of the Boscombe Sand.

A section in Warren Hill Sand in an old pit [SZ 169 9073] north-west of the Coastguard Station shows 7 m of thinly bedded and cross-bedded, buff, very fine-grained sand. CRB

Chama Sand

The Chama Sand is about 6 to 9 m thick between Crow [SU 163 040] and Burley [SU 212 030]. An auger hole [SU 1935 0339] north of Black Bush penetrated pale greenish grey, slightly clayey, silty fine-grained sand beneath flinty sandy head. Near Bagnum Farm [SU 1793 0284], buff, clayey, fine-grained sand occupies a slope, with spring-lines separating it from the Becton Sand above and Barton Clay below. The base of the slope occupied by the Chama Sand forms a concave-upward feature as it passes down to the gentler wet clayey slope below.

North of Burley Park, Chama Sand caps small hills on the Barton Clay. On one [SZ 2150 0414], greenish grey clayey sand was augered. Near Burley Street [SZ 2194 0346], 1.6 m of clayey, greenish grey sand passing down into dark grey silty clay, underlies 1 m of Head. In a pit [SZ 1992 9782] east of Bransgore, yellowish orange, slightly clayey, silty, fine-grained sand underlies river terrace gravel. A borehole [SZ 2020 9764] near Beech House, Bransgore, proved 1.4 m of pale grey clayey sand to sandy clay underlying 4 m of river terrace gravel. About 1 km south-east, another borehole [SZ 2064 9681] showed 3.5 m of very silty, mottled orange and yellow, fine-grained sand below 3.3 m of river terrace gravel. A third borehole [SZ 2084 9602] proved sandy blue clay with shells, under 5.8 m of gravel. An auger hole [SZ 2129 9540] in the stream bed south of Hinton showed greenish grey clayey, silty, fine-grained, weakly glauconitic sand. ECF

Grey, clayey, fine-grained sand occurs in the railway cutting [SZ 2142 9480; 2180 9475] south-east of Hinton. An auger hole in the stream bed 150 m north of the cutting proved grey-buff, clayey, fine-grained sand beneath 0.6 m of gravel. Pale grey, clayey, fine-grained sand occurs in the bed of the stream [SZ 2207 9428] which flows through Walkford.

There are poor exposures in up to 8 m of Chama Sand along the cliffs from south of Naish Farm [SZ 2240 9317] to the edge of the district, but much of the section is obscured. One exposure [SZ 2243 9317] showed 1.5 m of gravel overlying 1 m of very fine-grained, orange and buff, silty sand (Becton Sand), passing down into grey, very clayey, fine-grained sand.

On the west side of Chewton Bunny, pale grey, clayey, fine-grained sand was augered [e.g. SZ 2234 9402; 2237 9426]. CRB

Becton Sand

In the northern part of the district, the Becton Sand is about 7 m thick; it thickens southwards to 24 m on the coast.

Small exposures of fine- to very fine-grained sand are common on the high ground around Crow Hill Top [SU 180 039]. One [SU 1795 0390] showed over 3 m of yellow fine-grained sand with *Ophiomorpha* burrows.

On the west-facing slopes of Burley Hill, exposure is poor owing to a wash of gravel. A section [SU 1966 0194] in the old railway cutting exposed 6 m of fine-grained yellow sand beneath the Headon Formation. Farther west, in another cutting [SU 1795 0203], over 4 m of yellow, fine-grained sand underlies sandy gravel. South of the railway, a small pit [SU 1883 0133] showed the following section:

	Thickness m
Headon Formation	
Clay, buff to orange, extremely sandy, and clayey fine-grained sand	0.5
Becton Sand	
Sand, very slightly clayey, with some zones of ferruginous cement; molluscan casts in some cemented areas	1.2
Sand, yellow, fine-grained	0.2

A section [SU 2068 0336] between Burley Street and Burley exposes the junction with the Headon Formation and shows: extremely sandy clay and clayey fine-grained sand, with a thin buff ferruginous layer at the base, 1 m; on yellow, fine-grained Becton Sand, over 1 m. In Burley, an exposure [SU 2115 0302] shows 0.2 m of red-stained clayey sand overlying 2 m of fine-grained, clayey, yellow sand. On the plateau east of the village, a borehole [SU 2146 0281] showed over 3 m of fine-grained, yellow, buff and orange sand below 8.2 m of terrace gravel.

Becton Sand is exposed in a small pit [SZ 1914 9940] at Bransgore where gravel wash overlies fine-grained grey sand. A piston sampler hole [SZ 1929 9941] penetrated the following section:

	Thickness m
Gravel, orange-brown, clayey, sandy (? made ground)	1.95
Becton Sand	
Clay , orange-brown and grey, extremely sandy, with lateritic fragments	0.90
Sand, buff to brown, clay-free, fine-grained	1.00
Sand, orange to yellow, very fine-grained	1.25
Sand, orange to yellow, fine-grained	1.60

Another piston sampler hole [SZ 1948 9939] proved 1.2 m of orange to buff, fine-grained sand with rootlets, beneath about 10 m of Headon Formation. Nearby, test pits [around SZ 1960 9866] exposed several metres of buff, well-sorted, fine-grained sand.

Auger holes in the bottom of a small pit [SZ 2007 9861] on the south side of Thorney Hill proved 1.6 m of very fine-grained sand with buff, clayey, silty sand and very fine-grained buff sand just below the Headon Formation.

Buff, slightly clayey, very fine-grained sand, below carbonaceous clay and sand of the Headon Formation was found in a small tributary valley [SZ 2272 9557] to the Walkford Brook. The following section [SZ 2272 9502] alongside a track shows:

	Thickness m
River Terrace Deposits (Eleventh)	
Gravel, orange, sandy	1.5
Headon Formation	
Sand, very fine-grained, clayey, with some less clayey bands	1.0
Sand, fine-grained, silty, with some stiff clay bands	0.4
Clay, stiff, grey and greenish grey	1.5
Sand, very fine-grained, silty, pale grey to buff, with a little mica	0.8
Clay, chocolate-brown, silty	0.1
Becton Sand	
Sand, clean, fine-grained, buff, yellow and white	1.0
Sand, fine-grained, slightly clayey, with some clay-free layers and sporadic lateritic layers	2.5

ECF

Near Walkford, fine-grained, orange-brown sand was augered on the valley sides [SZ 2207 9434] and in the railway cutting [SZ 2193 9473].

A section showing the junction of the Becton Sand and Chama Sand at the western end of the cliffs is described on p.72. Fine-grained orange sand occurs along the sides of Chewton Bunny. CRB

Becton Bunny Member

In the coastal areas east of Naish Farm, the Becton Bunny Member occurs in the middle of the Becton Sand, but it has only a limited outcrop within the district.

In Chewton Bunny, east of Walkford, fine-grained sandy clay was augered beneath river terrace deposits [SZ 2265 9481]. North of this, a pit [SZ 2260 9506] exposed over 3 m of olive-grey, very sandy clay. The member is not seen farther north. ECF, CRB

SEVEN

Palaeogene: Headon Formation

STRATIGRAPHY

The Headon Formation, about 50 m thick, crops out around Burley in the north-east and in the high ground east of Bransgore. Much of the outcrop is obscured by gravel and gravel wash. The formation is divided into three members, but only the middle (Lyndhurst Member) is named.

The formation consists of pale greenish grey, relatively sand-free, locally shelly clays, and roughly laminated, very fine-grained sand, silt and clay. There is usually a carbonaceous silt, commonly associated with lignite, at the base; it overlies a palaeosol with rootlets at the top of the Becton Sand. The clays generally have a high kaolinite content. The sand and silt content of the clays is commonly low, and consists mainly of subangular quartz ranging in size from 10μ to 100μ; some marcasite is also locally present. The lowest member of the Headon Formation at Bransgore contains more sand than at outcrop in the New Forest, as do corresponding strata at Hordle Cliff near Barton on Sea, outside the district.

The Lyndhurst Member, 13 m thick, is mainly concealed by terrace gravels between Beckley [SZ 220 966] in the south and Holmsley Ridge in the north, but crops out north-west of Thorney Hill Holms [SU 207 004] and in Holmsley Inclosure [SU 225 004]. It consists of bioturbated, sandy, greenish and olive-grey clay, with fine-grained, commonly clayey and silty sand; however, some of the sands are clean and well sorted, and show a symmetrical grain-size distribution similar to those of the Becton Sand. Whereas thin-walled brackish to freshwater molluscs are common in the lowest and uppermost members of the Headon Formation, the Lyndhurst Member contains thick-shelled marine molluscs.

The uppermost member of the Headon Formation contains less sand than the lowest, and consists mainly of shelly greenish grey clay. In the highest parts of the succession there are traces of red-stained clays. Fine-grained shell debris is commonly abundant, either disseminated throughout the clay or concentrated in beds and laminae.

ECF

Conditions of deposition

The marine regression represented by the Becton Sand continued with the deposition of the lowest member of the Headon Formation. The freshwater and brackish fauna, the clay mineralogy, the presence of lignitic material and of a palaeosol developed on the underlying Becton Sand all suggest that this lowest member was deposited in a freshwater lagoon, behind a beach-barrier sand. The sand beds within the lowest Headon member are thought to have been derived from easterly flowing rivers that deposited fans of sand in the lagoon and caused a local lowering of salinity (Plint, 1983b). The silts and finer grained sands may represent either distal deposits from the easterly flowing rivers or, in part, marine sands derived from the south-east. Lenticular lamination (hummocky cross-stratification) commonly seen in the Headon Formation may be due to winnowing by wind-generated currents.

The bioturbated, greenish grey and olive-grey clays and clayey, fine-grained, shelly sands of the Lyndhurst Member indicate the last-known Palaeogene marine transgression in the district. Well-sorted, very fine-grained sands near the top of the Lyndhurst Member may be barrier sands, deposited during a regressive phase as marine shelf clay and sand deposition retreated eastwards. They were succeeded by the freshwater lagoonal sediments of the uppermost member of the Headon Formation.

ECF

Details

LOWEST HEADON MEMBER

On the west side of Burley Hill, gravel and sand hillwash obscures most of the formation, but fine-grained clayey sand and sandy clay, characteristic of the lower part of the formation, were augered in places. An exposure [SU 1975 0305] in a small gully showed the following section:

	Thickness m
Head	
Sand, clayey, grey, orange-stained, with patches of flints	0.8
Headon Formation	
Sand, slightly clayey, with disrupted clay-rich laminae; sporadic layers up to 5 mm thick of grey red-stained clay	0.5
Clay, pale grey, stiff, in layers up to 50 mm thick with 10–40 mm interbeds of fine-grained sand, some of which are red-stained	0.2
Becton Sand	
Sand, fine-grained, yellow, with some ferruginous layers and clay laminae	0.5

Exposures in the old railway cutting [SU 1966 0194] revealed:

	Thickness m
Head	
Sand, clayey, gravelly	0.5 to 1.0
Headon Formation	
Clay, greenish grey, with lateritic seams	0.5
Sand, fine-grained, clayey, bioturbated	1.5
Becton Sand	
Sand, fine-grained, yellow	6.0

A Dutch auger hole nearby [SU 1995 0198] penetrated 3.2 m of greenish grey stiff clay on over 0.1 m of laterite. Another [SU 1998 0193] penetrated: very sandy clay with laterite, 1.5 m; clayey, brown sand with laterite, 0.2 m; on fine-grained, clean Becton Sand, 0.4 m

South and south-east of Burley Hill, exposure is poor. An old pit [SU 2018 0260] exposes a small section in yellowish green, stiff clay with a few sand layers. A road cutting [SU 2068 0336] revealed over 1 m of extremely sandy buff clay and clayey sand, with a thin ferruginous layer at the base, resting on Becton Sand. South of Burley [SU 2095 0215], a borehole entered the Headon Formation below 5.6 m of river terrace gravel and proved: silty, yellow and grey clay, 1.9 m; on silty, fine-grained, quartzose, yellow sand, becoming grey below 8 m depth, 1.1 m. South-east of Burley, a borehole [SU 2267 0259] showed 3 m of silty yellowish brown to dark greenish grey clay below 3.1 m of gravel. Another borehole 750 m south-south-east [SU 2276 0180] encountered 3.2 m of yellow fine-grained sand, with a 0.5 m-bed of firm silty clay in the lower part, below 3.8 m of river terrace gravel.

The upper part of the lower Headon Formation on Dur Hill Down is mainly clay, passing down into clayey sand and extremely sandy clay, commonly with lateritic material at the base. A borehole [SU 1919 0125] shows the following succession, beneath 1.1 m of gravelly head: yellow and grey mottled clay with thin silty laminae, becoming more sandy below 3.4 m depth, 2.7 m; fine-grained, silty, yellow and buff sand, becoming pale grey below 4.1 m, 1 m.

A section [SU 1883 0133] on the east side of Bisterne Common showed 0.5 m of buff to orange, extremely sandy clay and clayey fine-grained sand resting on Becton Sand. Nearby [SU 1885 0136], there is much surface lateritic debris derived from the basal Headon Formation.

A borehole [SU 1866 0046] near Avon Tyrrell revealed 5.8 m of pale grey and yellow silty clay, passing down into silty, fine- grained sand at a depth of 6.1 m, under 1 m of gravel.

A piston sampler hole [SZ 1948 9939] on the west side of Thorney Hill showed the following section:

	Thickness m	*Depth* m
Headon Formation		
Clay, greenish grey, yellow-stained, sandy, with diffuse bands of silt	2.00	2.00
Sand, fine-grained, orange and grey, bioturbated, clayey	1.65	3.65
Clay, silty, greenish grey, with masses of very fine-grained sand and silt	0.35	4.00
Sand, fine-grained, and silt, buff, clayey, with thin clay beds	1.00	5.00
Clay, very sandy, orange, brown and grey, with silt pods and lateritic layers	1.00	6.00
Sand, very fine-grained, and silt, orange, with rough lamination at 7.2 to 7.55 m	2.05	8.05
Clay, orange-stained grey, passing down rapidly into olive-grey, bioturbated, extremely sandy clay	1.05	9.10
Clay, silty, dark greyish green, with silt beds and pods	0.55	9.65
Clay, dark brown, carbonaceous	c.0.35	c.10.00
Becton Sand		
Sand, fine-grained, orange to buff	1.20	11.20

Pits [around SZ 1960 9866] hereabouts showed greyish brown sandy clay overlying Becton Sand. East of Thorney Hill, auger holes proved yellow and greyish green clay; one [SZ 2024 9951] showed layers of silt within the clay. About 3 to 4 m above the base of the formation, an impersistent, fine-grained, slightly clayey, silty sand, 2 to 3 m thick, crops out around the valley sides. Springs are common at its base [e.g. SZ 2002 9940].

The Headon Formation is about 50 m thick east of Forest Lodge [SZ 216 984].

Exposures and auger holes [SZ 2281 9609] along the upper slopes of the Walkford Brook valley, proved 0.2 m of orange-mottled greenish grey sandy clay with silt layers, overlying 0.2 m of clayey fine-grained sand, above over 1 m of grey, very fine- grained running sand. Farther south, in a side valley [SZ 2273 9586], brown carbonaceous clay and sand, overlying Becton Sand were augered. The contact between the Headon Formation and the Becton Sand is described on p.73.

Lyndhurst Member

Fine-grained micaceous sand occurs beneath river terrace gravel in Holmsley Gravel Pit [SU 215 010] on Holmsley Ridge. In a ditch [SU 2165 0094], 0.7 m of fine-grained, yellowish orange sand, with a ferruginously cemented sandstone at the top, underlies clay of the upper Headon member. A similar sand occurs between 3.8 m and 6.8 m in a borehole [SU 2175 0111] nearby. Most sand samples from the Holmsley Gravel Pit show sorting and skewness characteristics similar to the shore-face sands of the Becton Sand (Figure 13H). South-east of the pit, auger holes in a stream bed [SU 2233 0016] revealed greyish green, very clayey, very shelly sand. Another auger hole [SU 2227 0006] contained very shelly, greenish grey, extremely sandy clay. A borehole [SZ 2258 9931] south-west of the last locality showed the following section, beneath 3.8 m of river terrace deposits:

	Thickness m	*Depth* m
Headon Formation		
Upper Headon member		
Clay, greenish grey, but yellow-weathered near the top and with some red staining, smooth to extremely sandy, with layers of silt; molluscan shell debris common	18.5	18.5
Lyndhurst Member		
Clay, sandy to very sandy, bluish grey to olive-grey, with some layers of sand and thick-shelled bivalves; rootlet bed at 25.7 – 26.3 m	2.4	20.9
Sand, dark grey, clayey, silty, medium-grained, with bands of greenish grey clay	2.3	23.2

Uppermost Headon member

Most auger holes east of Thorney Hill Holms [SU 208 003] yielded yellow-weathering, greenish grey, sticky, commonly shelly clay. Locally, some silty, clayey, fine-grained sand occurs [SZ 2151 9950; 2147 9967]. A borehole [SU 2131 0015] showed 1.3 m of laminated silty clay, overlying 2.6 m of silty clay with *Corbicula* sp., beneath 3.3 m of sandy gravel. Red-mottled, greenish yellow clay was noted locally in the highest parts of the succession [e.g. SZ 2385 9928].

Fine-grained clayey sand is exposed sporadically in the floor of a worked out gravel pit [SZ 2055 9885] near Hill Farm; and fine-grained greenish grey sand was augered [SZ 2130 9803] above greenish grey clay south of Plain Heath. Most samples seen between Beckley [SZ 222 967] and the edge of the district consist of fine-grained sand, usually clayey, but in many localities [e.g. SZ 2248 9673] stiff grey silty clay also occurs. In one auger hole in a stream bed [SZ 2284 9716] south of Ossemsley, dark olive-grey shelly clay was found.

The uppermost Headon member was proved in many boreholes in the Plain Heath [SZ 218 990] to New Milton area (Clarke, 1981). Auger holes in the valley [SZ 225 989] north of the Walkford Brook showed mainly sticky, yellow-weathering, greenish grey, commonly shelly clay.

A borehole [SZ 2087 9883] south-east of Thorney Hill proved the uppermost Headon member, below 4.5 m of river terrace gravel, as follows: fine-grained, yellow sand, 1.1 m; silty, mottled orange and brown clay, becoming dark greenish grey below 7.1 m depth, and with molluscs at 8.8 m depth, 2.1 m. Farther east, another borehole [SZ 2371 9873] encountered 1.7 m of greenish grey shelly clay below 4.8 m of river terrace gravel. A third borehole [SZ 2242 9755] showed 2.2 m of silty, laminated, orange and yellow-mottled greyish brown, in part lignitic clay, below 3.6 m of river terrace gravel. ECF

EIGHT

Structure

The district is structurally part of the Wessex Basin of southern England (Figure 19), the tectonic evolution of which was dominated by structures present in the underlying Variscan basement (Chadwick, 1985; 1986). The basement comprises predominantly east–west-striking, folded and cleaved rocks traversed by major thrust fault zones and

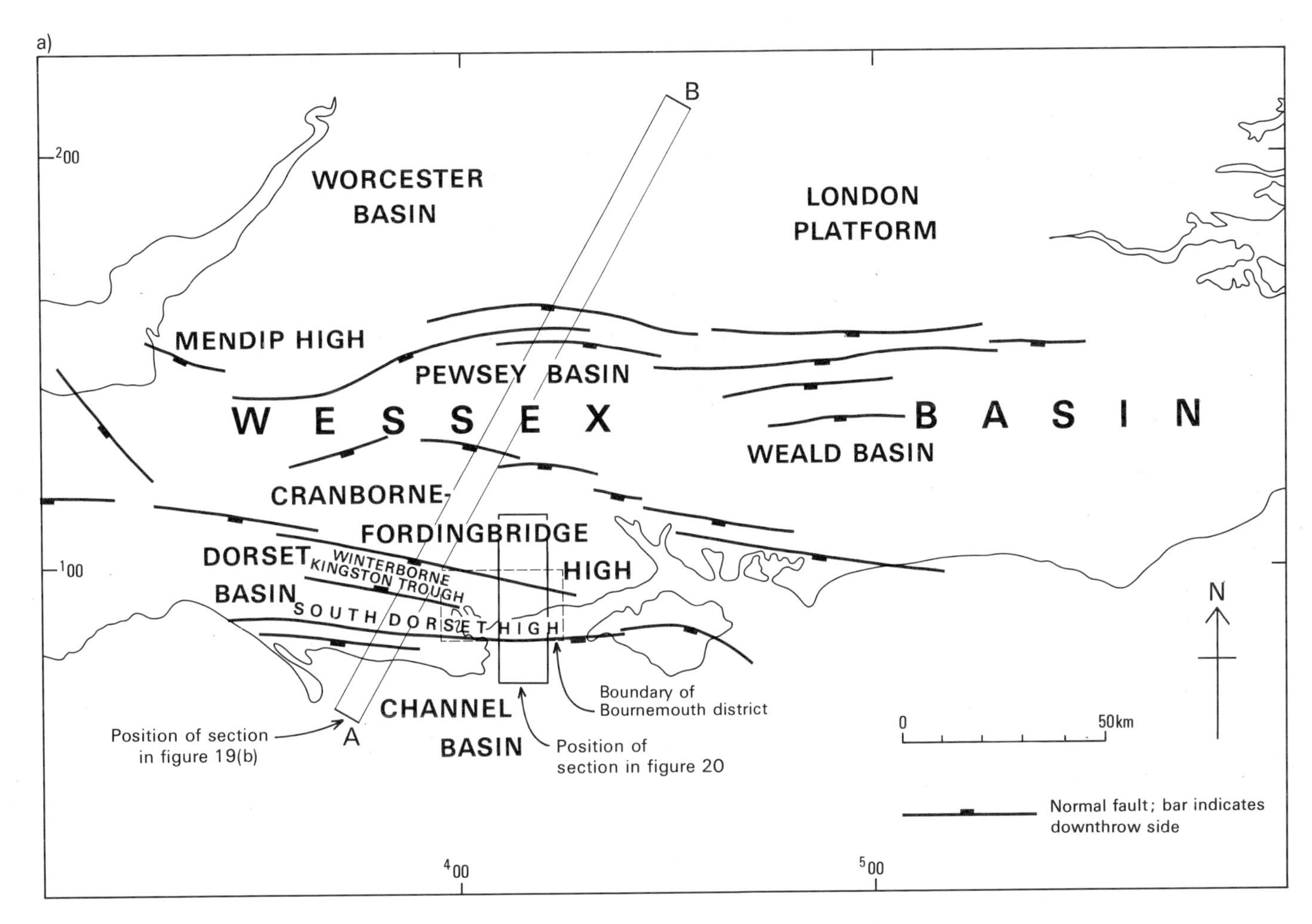

Figure 19a Permian to Cretaceous structural provinces in southern England (based on Chadwick, 1986, fig. 4a).

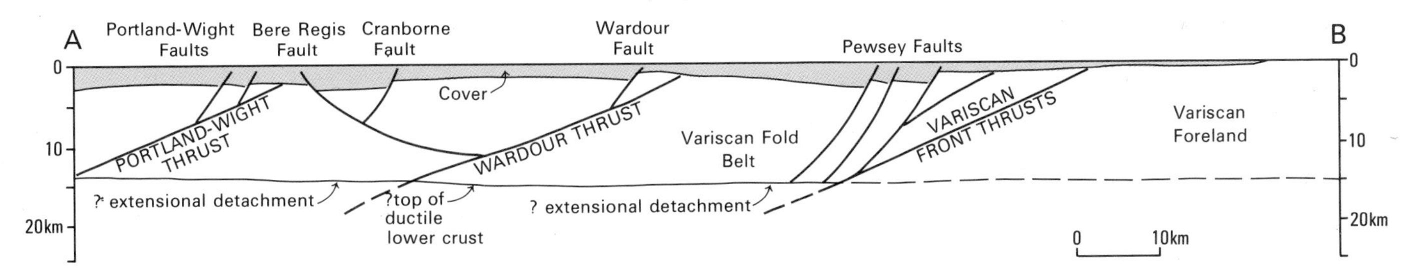

Figure 19b Deep crustal sketch-section across the Wessex Basin (based on Chadwick, 1986, fig. 8). See Figure 19a for line of section.

minor north-westerly trending wrench faults. Early and mid-Mesozoic extension reactivated the thrust faults to form a series of east–west-trending graben and tilt-blocks defined by syndepositional, generally south-dipping, normal faults. Subsequent post-Aptian, regional, unfaulted subsidence gave way to Cenozoic inversion, which reactivated some of the earlier faults to form generally north-facing, monoclinal flexures and, in places, reverse faults. At the same time, the general shape of the Tertiary Hampshire Basin was imparted.

VARISCAN STRUCTURES

The concealed Variscides are represented by a featureless zone on seismic reflection profiles; their detailed structure is unknown in the district. Phyllites at Wytch Farm (see p. 6) have radiometric ages of 337 ± 5 to 357 ± 5 million years, which indicate mid to late Devonian metamorphism; no other structural information is available. The tectonic development of post-Variscan faulting demands, according to Chadwick (1986), that faults associated with the Wytch Farm Oilfield are synthetic to a reactivated, southward-dipping, low angle, Variscan thrust fault (Portland–Wight Thrust), soleing out near the top of the middle crust at a depth of about 15 km, and which must subcrop to the north of the oilfield (Figure 19b). Similar arguments require a comparable thrust to the north, above which the Winterborne Kingston Trough is developed between an antithetic northward-dipping Bere Regis Fault, and a complementary Cranborne Fault secondarily antithetic to the Bere Regis Fault (Figure 19b). These major putative thrust faults have not yet been detailed by seismic profiling.

POST-VARISCAN STRUCTURES

The principal feature of post-Variscan structure in the district is the contrast in tectonic style between rocks below and above the local base of the Cretaceous. The former were strongly affected generally by south-dipping, major normal faults developed in an extensional tectonic regime, whereas the latter were very gently folded or flexured, and traversed by minor faults generated by compressive movements. This reversal of structural style, or tectonic inversion, may be seen by the way the synclines and anticlines of the gently folded lowest Cretaceous (Figure 22) and the base Tertiary surfaces (Figure 24) approximately overlie the subjacent upthrown and downthrown areas respectively of the pre-Cretaceous rocks (Figure 21).

Pre-Cretaceous structures

The ?Permian to Jurassic strata are disposed in a broad west-north-west-plunging downwarp, the Winterborne Kingston Trough and its extension towards the south-east (Chapter 2; Figure 5). It is flanked to the north and south by similarly plunging, broad anticlinal areas, the Cranborne–Fordingbridge High and the South Dorset High respectively. In detail, these structural units are complexes of tilted fault blocks, and the dominant structures of the ?Permian to Jurassic rocks are the east–west to ESE–WNW-trending normal faults. The faults dip at 60° to 70°, probably flattening at depth, and have downthrows to the south or north (Figure 20). As a result of their (probable) early to mid-Mesozoic growth and Cenozoic reversal, the throws of the faults change with depth. None, however, displaces Cretaceous or younger strata within the district. From north-east to south-west the more important faults are (Figure 5):

Burley Fault

The Burley Fault, an east-south-east-trending normal fault, of north-easterly downthrow, enters the district south-east of Burley [SU 210 030] and dies out some 4 km to the north-west. Its throw increases towards the south-east beyond the district, where it may be traced to the Isle of Wight. Within the district, it throws Corallian Beds to the south against Kimmeridge Clay, beneath overstepping Cretaceous beds. The throw increases with depth, displacing Middle Jurassic strata some 75 m, the base of the Lias some 175 m, and the top of the Variscan Basement some 180 m down to the north.

Cranborne Fault

This, the most prominent of the faults, enters the district to the north of Shapwick [ST 942 020], whence it trends just south of east to near Barnsfield Heath [SU 130 000], beyond which it swings to east-south-east, before trending south-easterly to the east of Bransgore [SZ 200 980]. The last change of direction suggests the existence of a NW–SE-trending fault in the Variscan Basement. In the south-eastern part of the district, the Cranborne Fault throws down Oxford Clay and Corallian Beds to the south-west against Kimmeridge Clay beneath the Cretaceous unconformity. Its throw increases to the west and with depth. At the level of the Cornbrash, its displacement is some 25 to 30 m, increasing to about 75 m at the top of the Lias; the Variscan Basement is displaced about 300 m in the east, but over 600 m in the west.

Bere Regis Fault

Where it enters the district north of Lytchett Matravers [SY 940 980], this northerly dipping normal fault trends north of east; farther east, seismic evidence suggests that it may trend south of east before dying out on the north-western outskirts of Bournemouth. Beneath the Cretaceous unconformity, it throws Oxford Clay against Kimmeridge Clay in the west and, as the throw diminishes eastwards, it displaces Corallian Beds against Kimmeridge Clay, before dying out within the Kimmeridge Clay subcrop. The throw ranges from zero in the east to 75 m in the west at the level of the Cornbrash, increasing to 250 m at the top of the Variscan Basement.

Wytch Farm faults

These east–west-trending normal faults, located beneath Poole Harbour, form part of the Isle of Purbeck Fault Zone, which delimits the southern margin of the South Dorset High and which extends eastwards to form the Isle of Wight Fault Zone (Chadwick, 1986; Figure 3a). The throws of the faults, most of which dip to the south, are generally less than 100 m, but they are sufficient to repeat the Cretaceous subcrop of the Corallian Beds and Oxford Clay. Despite the relatively small throws, they are of great significance in

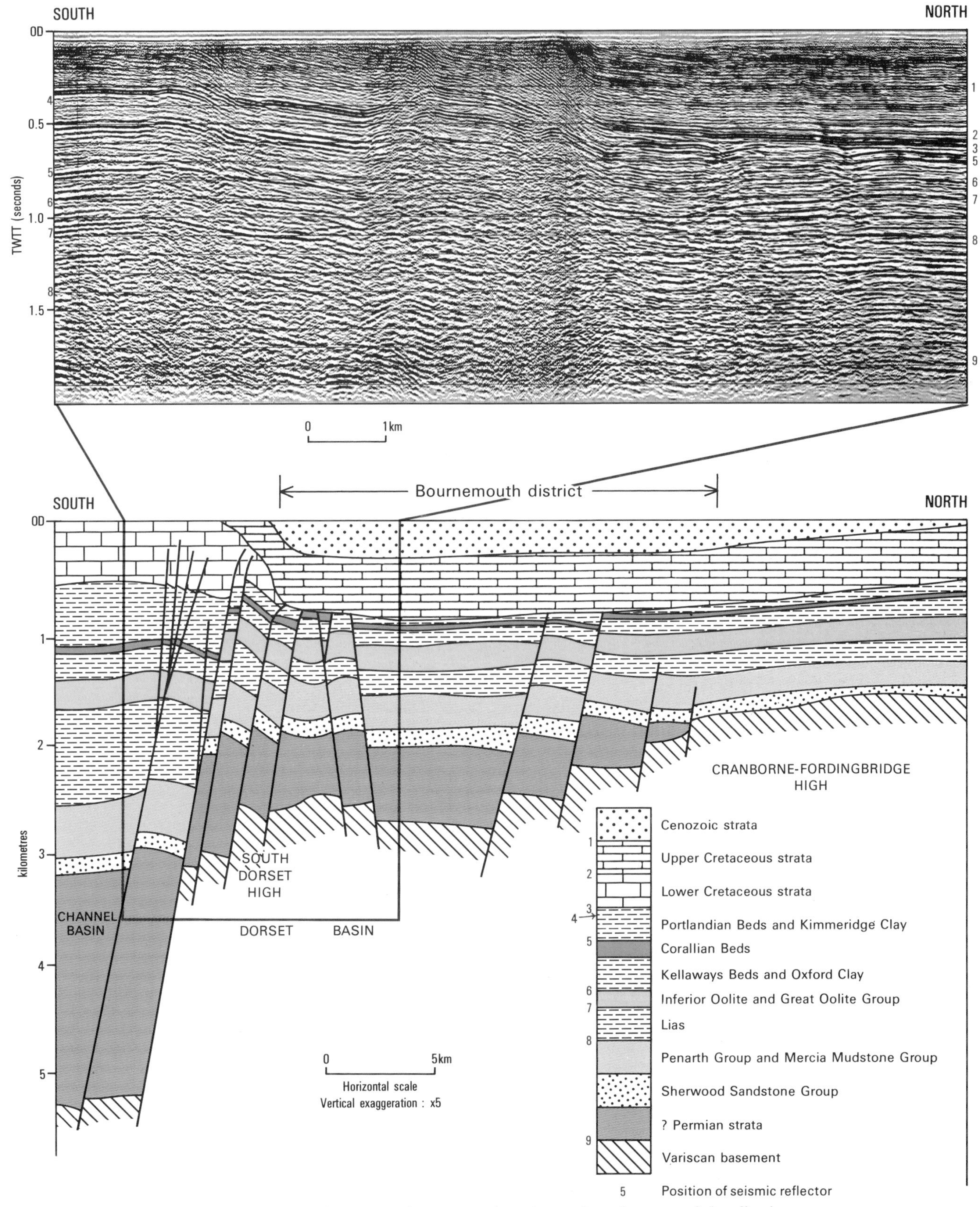

Figure 20 Simplified and generalised structural cross-section situated to the east of the district. Precise location is not shown for commercial reasons. Illustrated seismic section by courtesy of GSI Limited. Based on Penn et al. (1987, figs. 4, 5). See Figure 19a for approximate location of section.

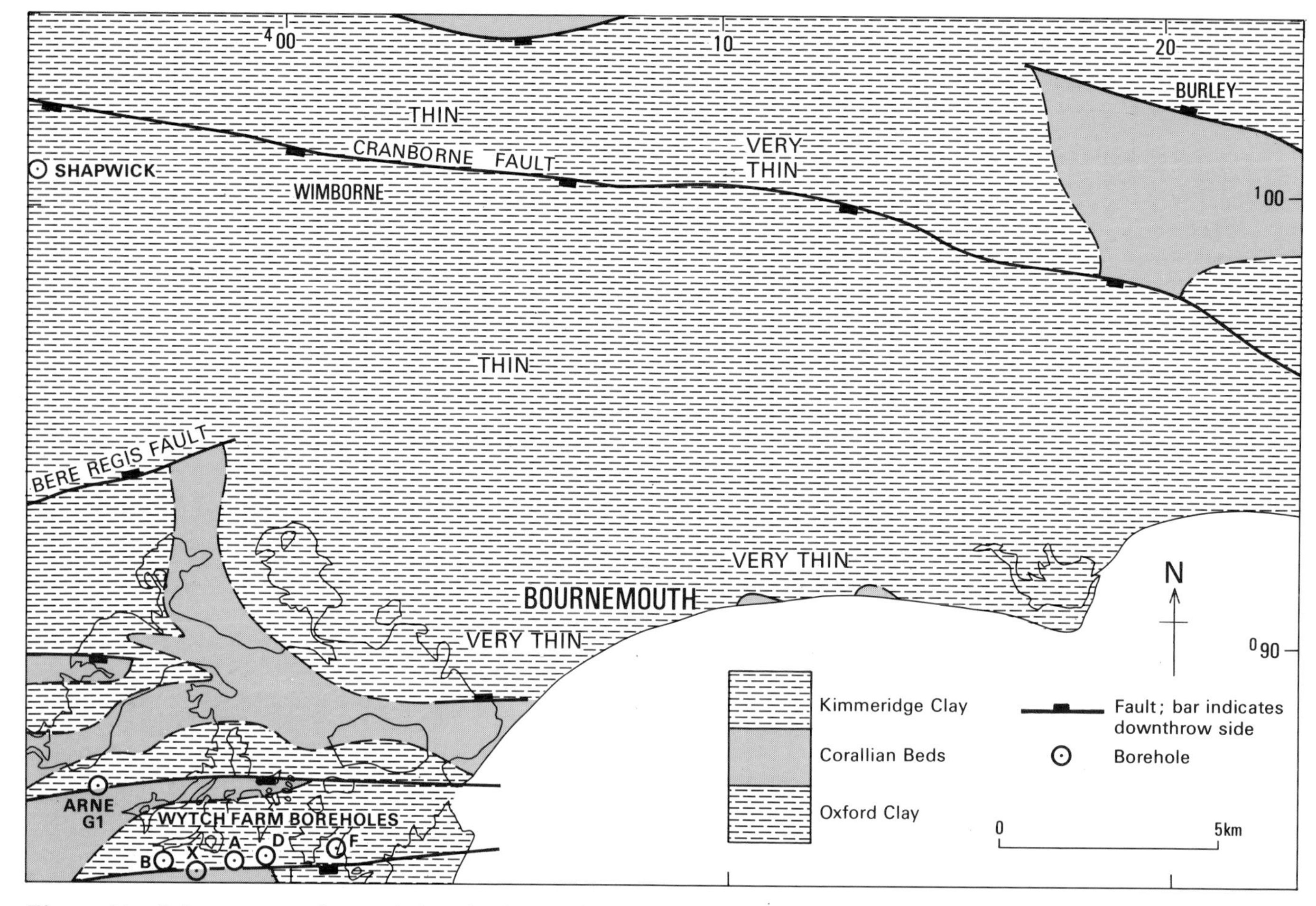

Figure 21 Subcrop map of strata below the Lower Greensand/Gault of the district.

hydrocarbon entrapment, because, by downfaulting superjacent mudstones and shales against subjacent reservoir rocks, they provide faulted closure. The southernmost fault, in particular, seals the southern margin of both Sherwood Sandstone and Bridport Sands reservoirs in the Wytch Farm Oilfield. The northernmost fault downthrows to the north and forms a northern seal to the Sherwood Sandstone reservoir.

Post-Cretaceous structures

The district lies on the western flank of the Hampshire Basin, where the dip of the Palaeogene strata is generally less than 1°SE towards the centre of the basin in the Isle of Wight. Variations in dip are caused by small, commonly asymmetrical, folds or flexures, which generally trend either east–west or NW–SE. The flexures commonly lie above folds affecting the basal Palaeogene (Figure 23) and basal Lower Greensand/Gault (Figure 21) surfaces. Some overlie faults in pre-Albian strata. Minor NW–SE faulting affects the concealed basal Cretaceous boundary, but of these faults, only the Christchurch Fault persists upwards to displace Palaeogene rocks, by a maximum of 30 m. The asymmetric easterly trending syncline through Poole Harbour overlies the Wytch Farm fault zone, and the complex synclinal structure north of Corfe Mullen [SY 978 980] is related to the Bere Regis Fault at depth. Like the underlying fault, the latter syncline dies out to the south-east. To the south-east, however, an easterly trending syncline through Christchurch Harbour suggests the presence of an easterly trending structure at depth. The broad south-east-trending anticlinal flexure passing south-west of Wimborne and north of Christchurch is the inverted image of the underlying Winterborne Kingston Trough. Its complementary synclinal tract similarly bestrides the underlying Cranborne–Fordingbridge High.

It is probable that many of the changes in thickness within the Palaeogene succession are controlled by contemporaneous movement of the deep-seated faults. For example, the thick and nearly complete Poole Formation in the structural downwarp under Poole Harbour thins abruptly north of an area of complex flexuring in the post-Albian rocks overlying pre-Albian faults. Farther north, the sequence below the Broadstone Clay and its underlying sand is missing, and the sand rests directly on London Clay. Similar syndepositional development of Cenozoic structure in response to movements along basement faults has been inferred elsewhere in the Hampshire Basin (Daley and Edwards, 1971; Plint, 1982; Edwards and Freshney, 1987b). ECF, IEP

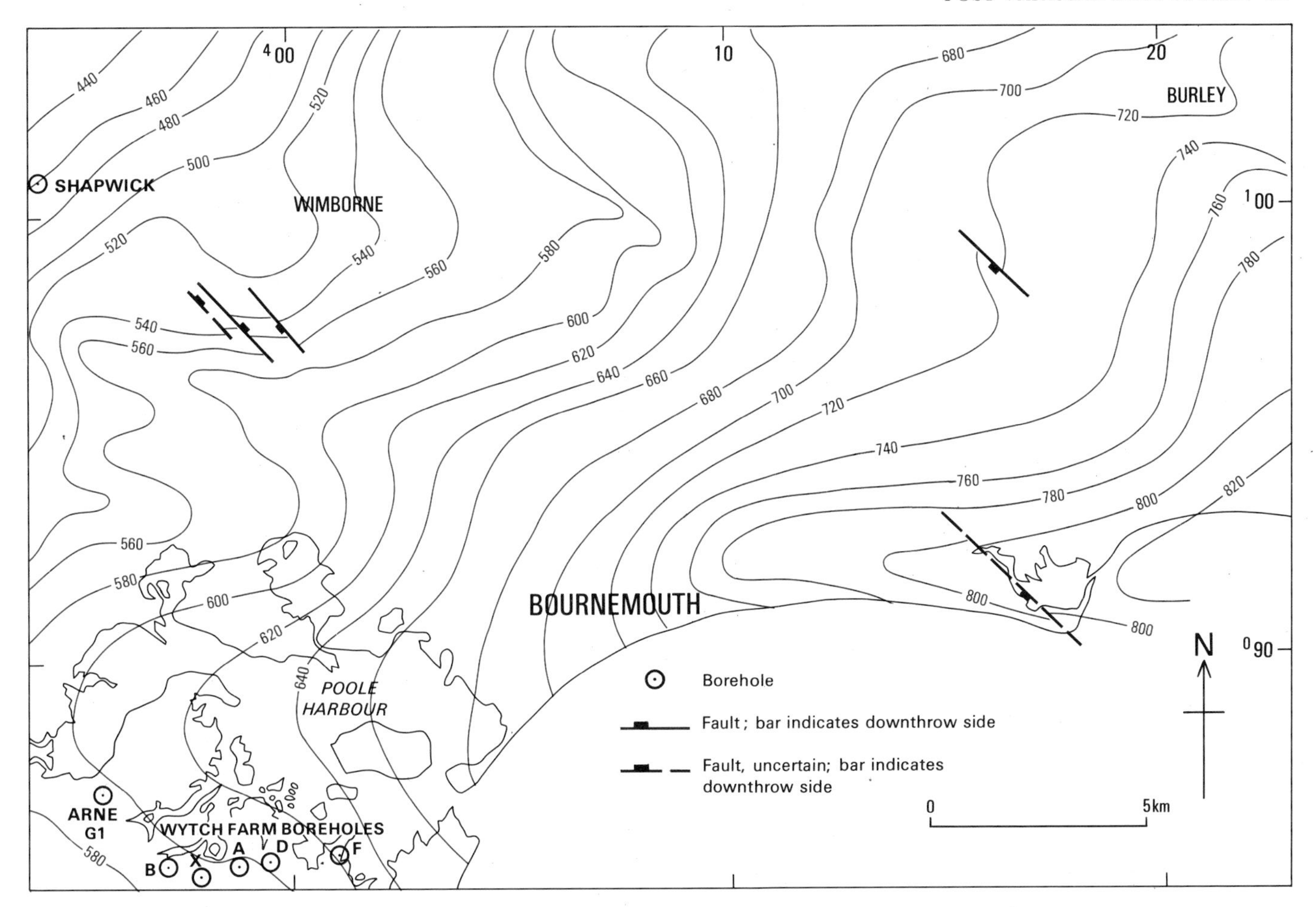

Figure 22 Structure contour map of base of the Lower Greensand/Gault of the district, based on seismic data (contours at 20 m intervals below OD).

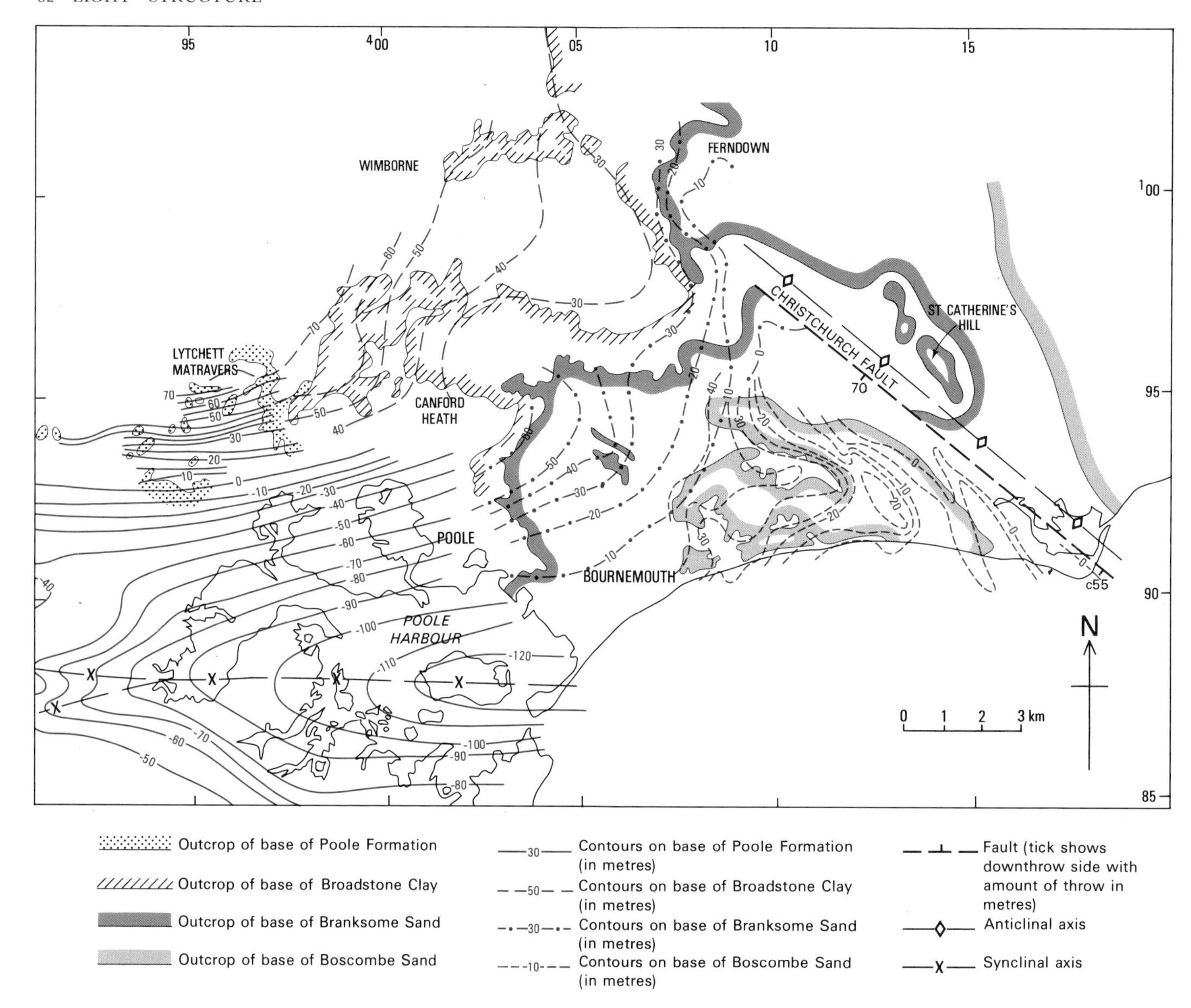

Figure 23 Contour map of the bases of the Poole Formation, Broadstone Clay, Branksome Sand and Boscombe Sand in the district.

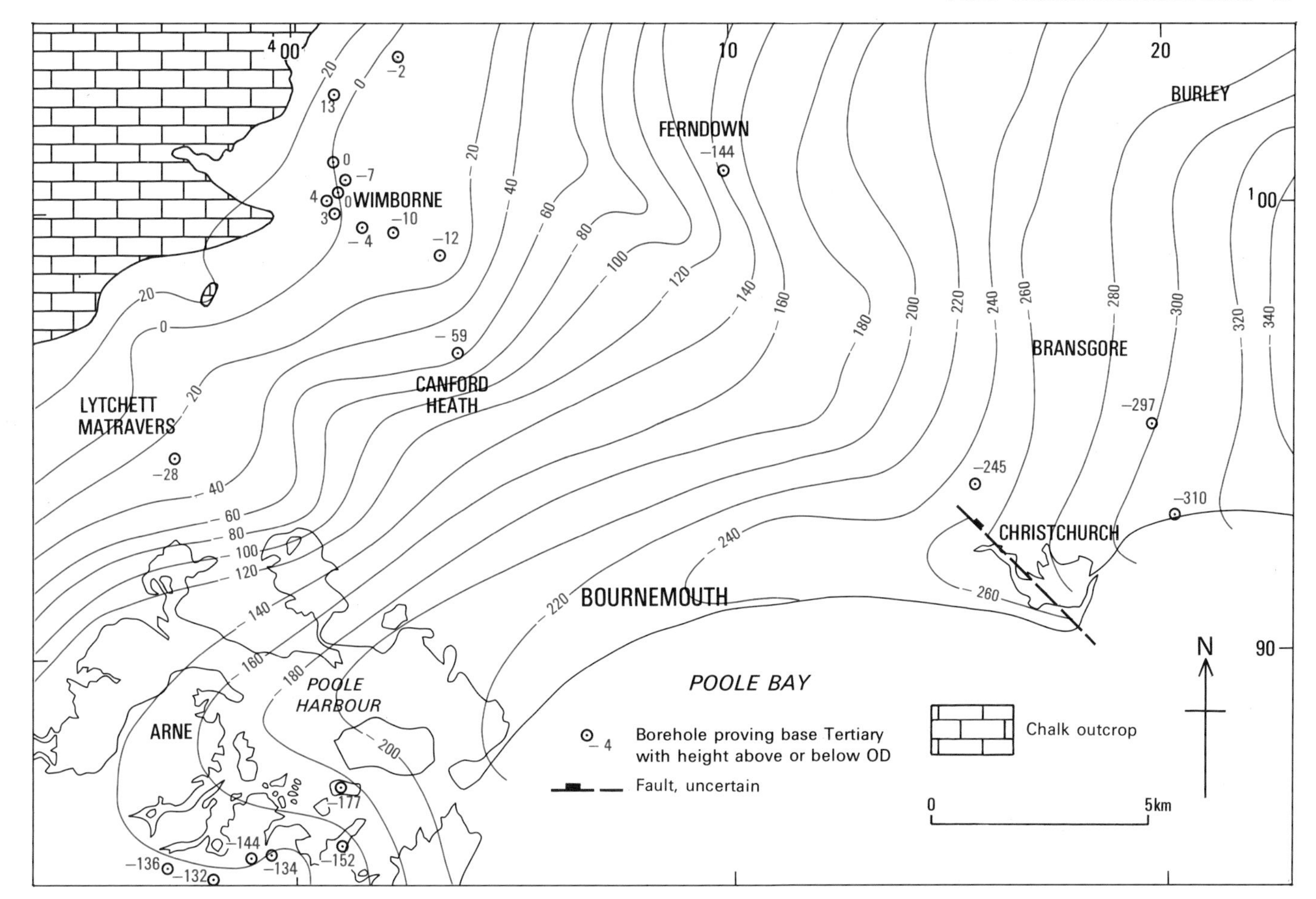

Figure 24 Structure contours on base of Palaeogene strata, based on outcrop, borehole and seismic data (contours at 20 m intervals above or below OD).

NINE

Quaternary

The Quaternary era saw many oscillations of climate, ranging from cold periglacial to warm temperate. During the more extreme cold periods, extensive ice-sheets pushed southwards across England, but did not reach the Bournemouth district. Marked changes in sea level were associated with these climatic oscillations. Meltwater rivers from the ice sheets or local impermanent ice caps were graded to base levels that were at times considerably above or below that existing at present. These rivers carried large volumes of sand and gravel which were deposited as a whole suite of fluvial deposits preserved today as the Older River Gravels and River Terrace Deposits. Periglacial conditions, which extended southwards beyond the ice sheet, were responsible for the formation of the Head Gravel and Head deposits. Because of the lack of fossils or radiometrically dated material it has not been possible to relate these older Quaternary deposits to the sequence of named stages recognised elsewhere in the British Isles.

During the Flandrian, the most recent of the Quaternary stages, extensive deposits of Alluvium and coastal spreads of Estuarine Alluvium, together with minor occurrences of Peat, Storm Gravel Beach Deposits and Blown Sand, were laid down.

HEAD

Head comprises heterogeneous deposits derived from downhill movement by solifluction of weathered surface material from the solid formations and older drift deposits. Much of the Head accumulated under periglacial conditions, when freeze–thaw activity disrupted strata which, during the spring and summer thaws, became sufficiently water-saturated and mobile to move downslope. Soil creep and minor mass movements have added their contribution and probably continue to do so. Some wind-blown material has possibly also been incorporated into these deposits.

Head deposits in the Poole–Bournemouth district occur in the valley bottoms and, to a lesser extent, on the valley slopes. Valley-bottom Head probably includes some waterlain material. Valley slope deposits, which commonly form sheet-like spreads, particularly on south-facing slopes, formed initially from a soliflucted downwash of gravel, sand and clay derived from the terrace deposits, and sand and clay derived from the solid formations; they were later augmented by colluvial hillwash material. Green (1947) used the term 'Bluff deposits' for the soliflucted material that occurs on the slope between terraces. Most of the flint implements found in the district have come from these sediments (see Calkin and Green, 1949). Head deposits also occur downslope from spring-lines at the base of many of the major sand units. These deposits are largely of colluvial origin, although they probably incorporate some soliflucted material.

With such a variety of source material, Head deposits vary markedly in their composition and thickness. Although individual deposits are commonly heterogeneous, their lithology generally corresponds to the upslope source material. In general, the thickness does not exceed 3 m, but up to 6.5 m of gravelly sand occurs on the slope between the Eighth and Fourth River Terraces between Charminster [SZ 100 946] and Littledown [SZ 128 930] (Freshney et al., 1984). In some cases it is impossible to distinguish it from its source deposits; similar difficulty is experienced in separating Head from weathered in-situ deposits.

Organic material and thin beds of peat occur in the Head; they locally give rise to large boggy tracts such as The Bog [SU 0480 0025] and Bagnum Bog [SU 182 025]. They are particularly associated with Head deposits related to spring-lines, and are especially common in the headwaters of the River Avon (Freshney and Bristow, 1987) and on the Arne Peninsula (Bristow and Freshney, 1987).

Head consisting mostly of sand derived from blown sand deposits, some of which is rich in organic matter and locally overlain by peat and peaty clay, occurs on the Studland Peninsula. There, the Head is underlain at shallow depth by Oakdale Clay and forms marshy ground, in part covered by lakes.

On Hengistbury Head, a small fan-shaped, largely unvegetated area of modern colluvial material, mapped as Head, consists dominantly of sand. It formed during periods of heavy rain when sheet flooding of sand-charged water crossed an area of made ground and debouched onto the flat surface of the river terrace deposits; the process continues today. Deposition is sufficiently vigorous to stifle all but the hardiest or most rapidly growing plants (Freshney et al., 1984).

Full details of the Head deposits are to be found in the Open-File Reports which cover the district (p.ix). ECF, CRB

HEAD GRAVEL

A small area of Head Gravel associated with Terrace 13 occurs near Burley [SZ 223 032], where it forms a north-facing apron below the scarp at the terrace edge. It consists of fairly loose sandy gravel, but locally it is cemented by ferruginous material. ECF

OLDER RIVER GRAVELS

In the north-east, there is a fragmented spread of commonly cryoturbated orange-brown, sandy, clayey, gravel of Terrace 2 of the Older River Gravels (Edwards and Freshney, 1987b). The gravels are higher than those of terraces related to the present-day river systems and are probably the deposits of Pleistocene sheet floods. Flint implements of

Acheulian age have been found on their surface outside the district. ECF

RIVER TERRACE DEPOSITS

Extensive river terrace deposits laid down mainly by the rivers Avon and Stour, together with lesser spreads along the rivers Allen, Uddens Water, Moors, Sherford, Piddle and Frome, occur at fourteen levels within the district (Figures 25 and 26). They range in height from 0.5 m to 61 m above their respective alluvial plains. The higher terraces in the south and east, from the Eleventh to the Fourteenth, were probably associated with the proto-Solent drainage system (Reid, 1902). The Ninth and Tenth Terrace Deposits were deposited at about the time the Stour and Avon valley systems were being established, and they cannot reliably be related to either the proto-Solent or the modern rivers. The First to Eighth River Terrace Deposits bordering the Stour and Avon are directly related to these present-day rivers.

During the original geological survey, the deposits now regarded as river terrace deposits were grouped as 'Valley Gravels' or 'Plateau Gravels'. They correspond respectively to terraces 1 to 5 and 6 to 14 of this account. Although called Plateau Gravels, it is clear that these high-level spreads of sand and gravel were regarded as river terrace deposits; some, for example the Lambs Green Terrace, were individually named (White, 1917).

Bury (1933) divided the Plateau Gravels into Upper Plain and Lower Plain suites, the former corresponding approximately to terraces ?10 to 14, and the latter to terraces 6 to ?10 of this account. On the evidence of the flint implements, Bury (1933) concluded that the gravels of the Lower Plain were older than those of the Upper Plain.

Green (1946) recognised and named sand and gravel spreads at nine levels within the district. Whilst some of his names, for example the Christchurch Terrace, were of local origin, others were derived from supposed correlatives in Sussex, Devon and the Thames Basin. Some of Green's terraces, particularly the higher ones, correlate specifically with terraces of the numbered sequence of this account, but others appear to be composite. The latter can be split into parts, each corresponding to a numbered terrace (Table 2); for example, the Staverton Terrace corresponds in part to Terrace 4, and in part to Terrace 5. Sealy (1955) recognised eight terraces related to the River Avon. The numbered sequence adopted here differs from that of Clarke (1981) for the upper reaches of the Avon and Stour because of the recognition of lower terraces in the south (not seen by Clarke) and of a number of higher terraces, particularly in the Bransgore area (Table 2). The relative positions of the various terraces and their heights above the floodplain are shown in Figures 25 and 26.

The river terrace deposits consist mainly of flint gravel, which is commonly very sandy. There are small proportions of chert, 'sarsen' stone and limestone, the last probably of Jurassic age. The maximum pebble size is usually around 5 cm across. The pebbles are usually subangular to subrounded, but in places, if there is a nearby source of Tertiary gravel, such as at Hengistbury Head, the deposits contain a considerable proportion of reworked well-rounded pebbles. Terraces 3, 6, 8 and 10 to 11 in the higher reaches of the Uddens Water have abundant rounded flints, which are in places the dominant clasts.

Pebble counts by Clarke (1981, table 4) show between 9 and 30% of rounded flint pebbles in samples from terraces two to five, and eight (of the present numbering system). The total percentage of flints within the district varies from 96 to 100%, with 0 to 2% vein quartz, 1 to 4% sandstone and up to 1% 'others'. Keen (1980) calculated slightly different percentages for the gravels of the Ninth River Terrace Deposits: flint, 92.5 to 95.4%, quartz, 0.8 to 2.7%, greensand chert, 0.8 to 3.5%, and 'far travelled rocks', 1.1 to 3.9%.

The First to Fourth River Terrace sequences may include up to 1.5 m of silty clay to clayey fine-grained sand at the top. The thickness of the river terrace deposits is highly variable and may be more than 8 m in the first and second terraces, and as little as 1 m in some of the higher terraces. An exceptional thickness of 11 m in the Eighth Terrace Deposits was encountered at one locality [SZ 0443 9728] south-west of Knighton. Most of the higher gravels, down to at least the third terrace, are cryoturbated, and typically show an irregular fabric characterized by involutions and flame structures.

Hand axes of several cultural types have been found in or on the surface of river terrace deposits, or in soliflucted gravel derived from them. The oldest implements, of Middle Acheulian age, are associated with Bury's (1933) Upper Plain gravels, and in particular his Sleight Terrace (Twelfth Terrace of this account) (Calkin and Green, 1949). These latter authors, however, point out that the majority of the Middle Acheulian implements come from the base of soliflucted gravel derived from the Sleight Terrace. Younger implements, of Upper Acheulian age, were found on the surface of this soliflucted spread (the 'Bluff-deposits' of Calkin and Green, 1949).

The Muscliff and Christchurch Terraces of Calkin and Green (1949), which correspond in part to terraces 2 to 4 of this account, yielded implements of Aurignacian (i.e. of uppermost Palaeolithic) type and, therefore, are probably of Devensian age. ECF, CRB

Details

RIVER TERRACES OF THE 'PROTO'-SOLENT

Fourteenth River Terrace Deposits

The outcrops of this terrace are limited to high ground in the east of the district. A borehole [SU 1866 0046] proved 1 m of silty sand on 3 m of sandy gravel, which in turn rested on the Headon Formation. A slurry pit [SU 1968 0011] showed over 2 m of orange, clayey, sandy gravel. A borehole [SU 2146 0281] near Burley revealed 3.8 m of clayey, sandy gravel, on 2 m of sandy silt, on 2.4 m of sandy gravel which rested on Becton Sand. ECF

Thirteenth River Terrace Deposits

These deposits are well exposed on the south side of Henbury Sand and Gravel pit. There, they vary generally between 1 and 1.5 m thick, but locally are up to 4 m thick. One section [SY 9606 9724] showed a lower unit, 1 m thick, of roughly bedded gravel, beneath 1 m of cryoturbated gravel. In another section [SY 9597 9111], 2 m of very clayey gravel are strongly cryoturbated. An old gravel pit

Table 2 Correlation of the terraces in the district

White (1917)	Bury (1933)	Green (1946)	Sealy (1955)	Clarke (1981)	This report
		Christchurch (part) (Hampreston area)			ALLUVIUM
				not present in Clarke's area	1
Valley		Staverton (part) Christchurch (part)		1 2 (part-Hurn area)	2
Gravels		Christchurch (part)	I	2, 3 (part-Hurn area)	3
		Staverton (part)		3 (part) 4 (part-Hurn area)	4
		Muscliff Christchurch (part) Staverton (part)		3 (part) + 4	5
Berry Hill		not present in Green's area	Not present in Sealy's area	5 (part-Berry Hill area)	6
		Second Lower Taplow			7
Bransgore Terrace (in east) Lambs Green Terrace (in west)	Lower Plain	First Lower Taplow	II + III	5	8
Higher Merley Terrace or Palaeolithic		Upper Taplow	IV	7	9
	— — ? — — —	Boyn Hill	Not present in Sealy's area		10
Plateau Gravels		First Lower Taplow (Bransgore area)	V		11
Eolithic Terrace	Sleight Terrace Upper Plain	Sleight	VI	8	12
	Corfe Hills	Ambersham	VII		13
			VIII	10	14

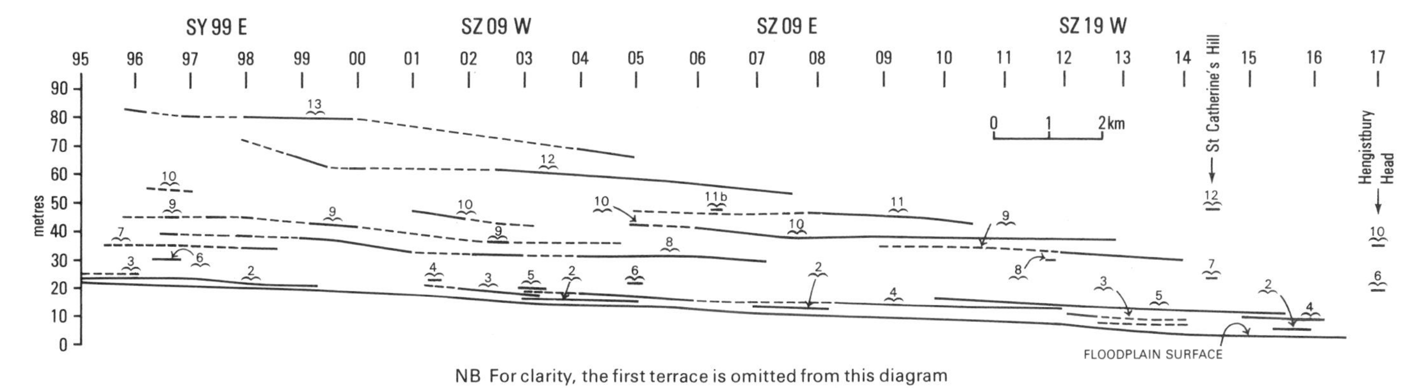

Figure 25 Profile of the River Stour and its terraces.

[SY 9604 9696] on the east side of Notting Hill exposed 2 m of unsorted and unbedded clayey gravel.

Boreholes at Corfe Mullen School show that the gravel varies from 2.8 m [SY 9970 9669] to 3.8 m in thickness [SY 9981 9665]. The deposits comprise dense, brown, sandy gravel with clay in the lower part, and up to 1.3 m of stiff, brown mottled, sandy clay or clayey sand in the upper part.

North of Broadstone, 1.2 m of cryoturbated gravel are exposed in a number of old pits [around SZ 000 971], and up to 1.8 m were formerly exposed in the railway cutting [SZ 003 973] 350 m to the north-east. To the south-east, there is an extensive shallow pit [SZ 0100 9625] in which gravel was worked to a depth of 3.6 m (Reid, MS map, 1894).

Boreholes [SZ 0378 9526; 0381 9526] on the east side of Canford Heath proved 0.5 and 4.1 m respectively of sandy clayey gravel. In a nearby partially filled pit [SZ 0395 9515], some 2 m of cryoturbated angular flint gravel, in an orange, medium- to coarse-grained sand matrix, overlies Parkstone Clay. 'Flames' of clay extend upwards for about 1 m into the gravel. CRB

A cutting [SZ 0475 9479] in the Ringwood Road showed: gravel, 0.2 m; orange sand and gravel, 1.6 m; sandy, white gravel, dark brown at the top, with an irregular cryoturbated base, 1.5 to 2 m; on sands of the Poole Formation, over 5 m. In the Newtown–Alderney area, drillers describe the gravel as angular and contained in a grey-brown, silty or clayey sand matrix, with thin bands of grey clay. CRB, ECF

The thickness ranges from 1.5 near Burley to 6.3 m [SU 2489 0143] just east of the district. An old pit [SZ 1965 9852], 400 m north-east of Bransgore House, shows over 3 m of ferruginous gravel with clayey sand layers. A section in Holmsley Gravel Pit [SU 2142 0106] shows:

	Thickness m
13th River Terrace Deposits	
Sand, gravelly, clayey, orange-grey and buff mottled; irregular cryoturbated, contact with gravel below	0.1 to 0.5
Gravel, sandy, orange-brown, cryoturbated with pebbles up to 5 cm across	2.5
Headon Formation	
Sand, fine-grained, slightly clayey, orange-yellow	1.0

Further south, near Thorney Hill, the deposits of the terrace [SZ 2087 9884] consist of 4.5 m of gravel overlain by 1.3 m of mottled yellow and orange sandy clay. ECF

Twelfth River Terrace Deposits

In pits east of Sleight [SY 99 98], Twelfth River Terrace Deposits have been dug extensively for sand and gravel. In the western pit, up to 1.2 m of medium gravel with interbeds of clayey, coarse-grained sand up to 0.3 m thick were seen [SY 9854 9802]. In another part of the pit [SY 9864 9814], up to 1.6 m of cryoturbated medium gravel with thin stringers of clayey, coarse-grained sand are exposed. Most clasts are less than 10 cm across and subangular, but some rounded pebbles occur.

The large sandpit [SZ 030 967] on the north of Canford Heath exposes several sections in up to 4 m [SZ 0272 9674] of sand and gravel. In another section [SZ 0304 9678], 3 m of sand and gravel, in units up to 0.3 m thick, are interbedded with thicker beds of sandy gravel. Cross-bedding in the sands is gently inclined to the east. CRB

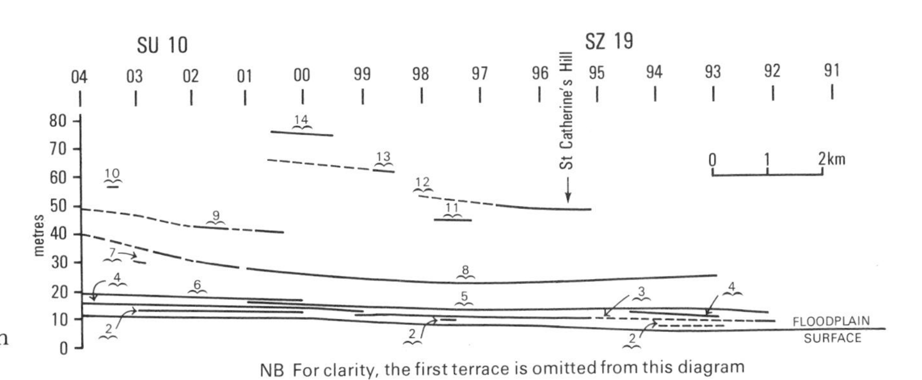

Figure 26 Profile of the River Avon and its terraces.

The maximum proved thickness at Talbot Heath is 5.1 m [SZ 0734 9349]. The deposits consist dominantly of angular, brown flint gravel in a matrix of clayey sand; interbeds of coarse-grained, locally cross-bedded sand up to 0.4 m thick occur.

At the eastern end of Talbot Heath, boreholes [SZ 0729 9357; 0734 9349] showed 3.6 and 5.1 m of gravel respectively. In another borehole [SZ 0721 9352], 4.1 m of brown sandy clay with much gravel overlie 0.5 m of gravelly coarse- and medium-grained sand. In other boreholes [e.g. SZ 0738 9361], the 5 m of terrace deposits are described as gravelly sand, overlying sandy gravel.

Near Wallisdown, boreholes [e.g. SZ 0545 9493] showed over 3.7 m of sandy gravel, and south of this, a borehole [SZ 0560 9424] penetrated 2.3 m of silty sand with scattered gravel resting on more than 0.2 m of dense sandy gravel. Near Talbot Heath, boreholes [around SZ 073 936] proved between 3.6 and 5.1 m of sandy gravel beneath up to 1.5 m of overburden. At the eastern end of the spread, boreholes indicate up to 4 m of gravelly sand and gravel. One [SZ 0751 9344], showed the following succession:

	Thickness m	*Depth* m
12th River Terrace Deposits		
Gravel, fine to coarse, very sandy, clayey in parts	1.50	1.50
Sand , fine-, medium- and coarse-grained, slightly silty, orange-brown, with fine, medium and coarse angular gravel	2.05	3.55
Branksome Sand	3.45	7.00

ECF, CRB

On St Catherine's Hill [SZ 144 955], the deposits are up to 2.8 m thick. There, a section [SZ 1432 9581] showed 1.95 m of fine, with some coarse, subrounded flint gravel, containing some small well-rounded flint and vein quartz pebbles, set in a matrix of dominantly medium-grained, subangular to subrounded, quartz and flint sand and silty sand. The deposit is patchily leached and generally cryoturbated. BJW

In the east of the district, south-west of Ossemsley, the thickness ranges from 2 to 5.5 m. A typical section is provided by the following borehole [SZ 2242 9755]:

	Thickness m	*Depth* m
12th River Terrace Deposits		
Clay, laminated, mottled grey and orange, with scattered flints below 0.8 m	1.0	1.0
Gravel, sandy, fine- to coarse-grained; subrounded to subangular flint, vein quartz and sandstone	2.6	3.6
Headon Formation	2.2	5.8

ECF

Eleventh River Terrace Deposits

At Gravel Hill near Merley, the road cutting [SZ 0165 9762] exposes 2 m of sandy medium gravel, above Broadstone Clay.

A section [SZ 0626 9340] in an old pit west of Talbot Heath reveals 2 m of ferruginously cemented, medium to coarse gravel. Boreholes [around SZ 061 935] at Rossmore show that most of the terrace deposits have been removed and the area partly backfilled.

Near Branksome, boreholes [SZ 0653 9150 and around 067 916] proved up to 3.8 m of dense, brown gravel and sand.

Boreholes [around SZ 0655 9160] at County Gates, Poole, proved between 3.3 and 3.7 m of coarse-grained, brown sand and flint gravel. Sand is the dominant lithology in the Wilderton Road area [around SZ 060 916], where up to 7.5 m of silty sand with fine and medium flinty gravel was recorded.

Boreholes [SZ 0707 9277; 0700 9264] at Winton proved 2 m of gravelly sand and very sandy gravel, and 3.3 m of sandy gravel respectively. A well [SZ 1023 9278] near Queen's Park Golf Course penetrated 0.6 m of made ground, resting on 2.74 m of gravel, on Boscombe Sand.

At Arne, a small pit [SY 9700 8813] at about 50 m OD exposed 2 m of brown-stained, subangular flint gravel. CRB

In the eastern part of the district, this terrace is 4 to 6.1 m thick. A borehole [SZ 2084 9602] in Hinton Park showed 0.91 m of stony brown clay, overlying 4.88 m of brown gravel, resting on Chama Sand.

Tenth River Terrace Deposits

A section [SZ 0403 9128] at Lower Parkstone showed 1.5 m of sandy gravel above Branksome Sand. Farther south, an old gravel pit [SZ 0417 9047] exposed 2 m of sandy gravel with a very clayey top, resting on Branksome Sand. A borehole [SZ 0804 9350] near Meyrick Park proved only 1.7 m of clayey, silty, gravelly sand, whereas in the eastern part of Winton [SZ 0908 9304], up to 2.3 m of sandy gravel occurs. In northern Westbourne, up to 2.9 m of sand, sandy gravel and sandy clay [SZ 0758 9130; 0764 9086; 0886 9076] have been proved.

Boreholes along the Wimborne Road [SZ 0895 9242 to 0883 9190] showed between 2.4 m and 3.5 m of gravelly sand and sandy gravel. Farther east, a borehole [SZ 0944 9187] proved about 4 m of gravel and sand. Nearer the cliffs, 4 m of dense, well-graded gravel and sand are overlain by 1.5 m of poorly graded, silty sand with some gravel [SZ 0994 9139].

At Springbourne, a borehole [SZ 1043 9256] penetrated 4 m of gravel with pebbles up to 50 mm, resting on Branksome Sand. In another borehole [SZ 1194 9205], 150 m to the east, only 0.7 m of silty, brown and orange sand with scattered gravel were proved, overlying 0.3 m of yellow sand with pockets of clay on Boscombe-Sand. ECF

Between 2 and 4 m of cryoturbated terrace gravels, covered by brickearth, are well exposed on the cliff top between Naish Farm [SZ 2250 9315] and the eastern margin of the district. CRB

Ninth River Terrace Deposits

A section [SZ 2134 9317] at Highcliffe shows 0.5 m of topsoil above 2 m of poorly sorted, poorly bedded sand and gravel. The top 0.3 m is cryoturbated and contains vertically aligned pebbles. Sections close to the feather edge of the deposits [SZ 2066 9312; 2234 9320] show strong cryoturbation, with pockets of gravel, 2 to 3 m deep, separated by tongues of Barton Clay rising up through the gravel to the surface.

TERRACES OF THE PIDDLE–FROME RIVER SYSTEM

Sixth River Terrace Deposits

At Arne, a section [SY 9735 8838] showed 0.6 m of gravelly sand above more than 1 m of gravel. On Brownsea Island, 1.5 m of gravel was seen [SZ 0183 8839].

Fifth River Terrace Deposits

An old gravel pit on Holton Heath [SY 9568 9107] exposed up to 1 m of orange-brown sandy gravel on sand of the Poole Formation. An exposure [SY 9847 9138] at Ham Hill in Hamworthy, showed 2 m of sandy gravel. ECF

South of Poole Harbour, scattered occurrences of Fifth River Terrace Deposits occur on the Arne and Fitzworth peninsulas where they have an upper surface at about 15 m OD. A section [SY

9605 8576] south-east of Slepe Farm shows 1.5 m of sand and gravel above Poole Formation sand. CRB

Fourth River Terrace Deposits

Near Black Hill, up to 1.8 m of sand and gravel overlie sand of the Poole Formation [SY 9499 9143]. CRB

An exposure [SY 9525 9130] on Holton Heath consists of orange-brown, coarse-grained sand with gravel and clay-rich layers, some of which are subvertical and red mottled. Another exposure [SY 9617 9098] showed 2.5 m of sandy gravel. Near Hamworthy, wells [about SY 9905 9112] proved up to 3.7 m of sand and gravel. A section at the top of an old pit [SY 9886 9122] exposed 2 m of clayey coarse- to very coarse-grained sand, locally very gravelly and cryoturbated. Another exposure in the old Lake Clay Pit [SY 9798 9092] revealed 1.25 m of sandy gravel containing masses of mottled sandy clay near the top. ECF

At Shipstal Point south of Poole Harbour, there are up to 3 m of sand and gravel at the top of the cliffs. One section [SY 9835 8827] showed 1.5 m of gravel overlying 1.5 m of sand with scattered flints. Another section [SY 9833 8824] showed 3 m of gravel.

Third River Terrace Deposits

South of the Sherford River, a small section [SY 9495 9206] showed 1 m of coarse-grained sand resting on 0.2 m of gravel, which overlies Oakdale Clay. Another section close by [SY 9492 9204] showed at least 1.5 m of coarse-grained sand. CRB

South of Poole Harbour, the surface of the deposits is about 5 m above OD. On Long Island, a cliff section [SY 9864 8809] showed 1 m of buff sand with scattered flints, resting on 1 m of cryoturbated, cross-bedded sand and gravel, which is channelled into Poole Formation sands. The deposits on the west side of Wytch Fir Pound consist of at least 1.2 m of mottled orange and grey, clayey sand above gravel. ECF, CRB

Second River Terrace Deposits

This terrace lies at a height of about 2 to 5 m above OD near Lytchett Minster and Hamworthy. Ditch sections [SY 9479 9275 to 9495 9266] alongside the A35 expose up to 1.5 m of clayey, coarse-grained sandy gravel. Further east, the deposit consists of at least 1.8 m of clayey fine-grained sand [SY 9636 9310], above more than 1.2 m of brown, sandy, clayey gravel [SY 9530 9261]. At Hamworthy, a borehole [SY 9987 9028] showed yellowish brown very sandy fine, medium and coarse gravel down to a depth of 3.2 m, with about 0.3 m of silty gravelly sand on top. Another borehole [SY 9922 9052] revealed 1.4 m of yellowish grey, very sandy gravel overlain by 1.1 m of black, gravelly, sandy silt.

First River Terrace Deposits

Under Old Poole [SZ 010 905], the upper surface of the terrace is between 1 and 2 m above OD. The thickness and lithology varies, with 0 to 3 m of organic sand overlying 1 to 3 m of gravelly sand or sandy clay with gravel (Freshney et al., 1985). ECF, CRB

TERRACES OF THE RIVER STOUR

Tenth River Terrace Deposits

A spread of gravel at Higher Merley, assigned to the Tenth Terrace, appears from its alignment to be related to the River Stour and not to the 'proto'-Solent. Possibly this spread is composite (Freshney et al., 1985). Boreholes [SZ 0098 9823; 0139 9792; 0161 9800] near Merley proved between 2.8 and 3.3 m of gravel consisting of angular to subangular brown flints, with clasts mostly less than 2 cm across, in a clayey, coarse-grained sand matrix. Boreholes at Oakley proved 0.9 m [SZ 0257 9806] to more than 2 m [SZ 0237 9809] of fine to coarse gravel in a silty or clayey sand matrix. North-west of Eastlands Farm, a borehole [SZ 0429 9668] proved 3.5 m of angular and subangular brown flint gravel in a sand matrix; most clasts were less than 5 cm across.

An exposure [SZ 0601 9603] at West Howe revealed 2.5 m of coarse flint gravel set in a sparse matrix of coarse- and medium-grained sand. A borehole [SZ 0758 9509] near East Howe passed through 2 m of made ground, on 1.1 m of sand and gravel, on Branksome Sand. BJW

Near Ensbury Park, a borehole [SZ 0749 9491] proved 2.5 m of compact sandy gravel; one farther east [SZ 0823 9496] showed 3.3 m of compact, silty, clayey gravel. A third borehole [SZ 0801 9427] penetrated 7.1 m of brown, very sandy gravel. ECF

Ninth River Terrace Deposits

These deposits, consisting of at least 1.5 m of sandy clay and clayey sandy gravel, have a fragmented outcrop. The most extensive spread is near Lambs Green in the western part of the district.

Eighth River Terrace Deposits

At Oakley, the thickness of these deposits varies between 2 m [SZ 0239 9848] and 4.6 m [SZ 0221 9842]. A borehole [SZ 0207 9870] proved 3.4 m of clayey sand and gravel, with pebbles mostly less than 1 cm across, but becoming coarser downwards with pebbles 3 to 4 cm across. A borehole [SZ 0315 9802] on the small spread of terrace south of Canford Park proved 1.3 m of very silty, slightly clayey sand and gravel; cobbles were recorded in the lower part. South-east of the park, a borehole [SZ 0358 9751] penetrated 4.7 m of very sandy, fine to coarse gravel, with cobbles in the lower part. Sections in the gravel pit south of Moortown Farm showed gravel thickening eastwards from 1.6 m [SZ 0403 9741] to more than 5.5 m [SZ 0434 9727]. At the latter locality, the gravel is divisible into an upper unit, 2.5 m thick, of brownish grey, clayey sand and gravel with the upper 1.5 m unbedded and the lower 1 m roughly bedded; resting on a lower unit of more than 3 m of roughly bedded, orange-buff sand and gravel. A nearby borehole [SZ 0443 9728] proved 11 m of angular and subangular flint gravel in a slightly clayey, coarse-grained sand matrix.

Near Knighton, more than 3 m of sandy and clayey gravel were proved in a borehole [SZ 0470 9663]. Farther east, between 2.4 and 2.9 m of clayey and sandy gravel were proved in boreholes. In one [SZ 0493 9647], the upper part consisted of 1.7 m of sand with scattered gravel clasts, above 0.7 m of sandy gravel.

Near Southbourne [SZ 1461 9154], a borehole showed 0.76 m of made ground overlying 2.13 m of gravel and sand, resting on Boscombe Sand. A borehole [SZ 1237 9332] at Porchester School penetrated 1.45 m of dark yellowish brown, sandy gravel, resting on Boscombe Sand.

Seventh River Terrace Deposits

Only scattered remnants of these deposits are preserved in the west, and no detail of their lithology is available.

Sixth River Terrace Deposits

In the western part of the district, sections along the old railway line [ST 9404 0037 to 9435 0003] expose up to 2.5 m of clayey gravel, gravelly sand or sandy clay. Two sections [ST 9407 0032; 9418 0020] showed up to 1.9 m of clayey gravelly sand resting on orange-brown sandy clay, but a borehole [ST 9423 0020] hereabouts proved 1.55 m of flint gravel and cobbles set in a matrix of firm brown clay.

North-west of Wimborne, along the River Allen, the deposits appear to be clayey angular gravel, whereas north-east of the town [SU 050 028], along the Uddens Water, there is a high proportion of well-rounded flints.

Fifth River Terrace Deposits

A borehole [SU 0447 0017] near Canford Crossing proved 3.5 m of gravel in a silty sand matrix resting on London Clay. Only one borehole [SZ 0328 9889] penetrated the deposits in the area between Canford Park [SZ 035 986] and Knighton [SZ 046 978], where they were described as 3.6 m of 'clay with flints'.

A borehole [SZ 1071 9499] south of Mill Throop proved, beneath 0.9 m of soil and sandy clayey silt, 2.55 m of dense to very dense, fine- to coarse-grained sand, together with fine, medium and coarse gravel resting on Branksome Sand. Another borehole [SZ 1454 9211] penetrated only 0.9 m of brown sand and roughly bedded gravel above Boscombe Sand. A third [SZ 1388 9205] encountered 3.66 m of yellow sand and fine gravel, overlain by 3.5 m of sand and clay, probably of solifluction origin.

Fourth Terrace Deposits

Boreholes at the Sewage works, Lambs Green, indicate a deposit of variable lithology. The westernmost borehole [SY 9944 9888] proved 2 m of very clayey sand and gravel. In another borehole [SY 9943 9892], 5.3 m of sand and sandy gravel were encountered.

Further east, near Wimborne, boreholes proved between 1.6 m of clayey gravel and sand [SU 0427 0000] and 3.15 m [SU 0419 0004] of silty sand and gravel in a clayey silt matrix. The extensive Fourth Terrace deposits south-east of Wimborne are divisible into two leaves, 4a and 4b. A borehole [SZ 0396 9983] in the lower leaf, Terrace 4a, proved 2.6 m of gravel in a matrix of sandy clayey silt or very sandy clay. A second borehole [SZ 0473 9986] in Terrace 4a encountered 1.5 m of clayey, silty sand and fine gravel, above 1.35 m of silty sand with abundant gravel. CRB

South of Muscliffe, a borehole [SZ 0980 9554] sited on Terrace 4b proved 0.9 m of topsoil, on 6 m of gravel, on Branksome Sand. The thickness of the terrace deposits between Hampreston and Longham varies from 4.8 m [SZ 0627 9782] to 3.1 m [SZ 0593 9882]. The gravels are described as coarse and sandy, with beds of sand. East of West Parley, the deposit varies from 2.8 m [SZ 0863 9744] to 5.6 m [SZ 0887 9860] thick. The gravel is predominantly coarse, and is set in a matrix of medium- to coarse-grained sand. South of the river, two boreholes [SZ 0406 9859; 0425 9797] proved thicknesses of 4.1 and 6.2 m respectively of angular to subangular brown flints, mostly less than 5 cm across, set in a sand matrix. In the second borehole, a 1 m-thick bed of coarse-grained, gritty sand occurs between the depths of 3 and 4 m.

An old pit [SZ 1045 9914] near Hurn Airport showed 0.2 m of brownish grey, flinty loam ('brickearth'), on 2.2 m of sandy fine gravel. A borehole [SZ 1083 9947] proved 0.5 m of topsoil and brickearth above 3.6 m of fine, with some coarse, subangular to subrounded, chiefly flint gravel, in a matrix of medium-grained sand.

The bluff between the two terrace leaves near Throop [SZ 111 955], although small, is clear. A borehole [SZ 1046 9520] south-west of Throop showed 0.8 m of made ground on 2 m of very clayey 'brickearth', resting on 2 m of fine and coarse, subangular to subrounded, mainly flint gravel, in a matrix of chiefly fine-grained sand; the brickearth in this borehole is unusually thick. Patches of silty sand and clay, usually less than 1 m thick and commonly with scattered patinated subangular flints at the surface, have been noted in several places in the area, but have not been mapped separately from the gravels. BJW

Boreholes in Christchurch [around SZ 158 929] proved thicknesses of between 2 and 3.3 m of sand and gravel.

Third Terrace Deposits

A borehole [ST 9499 0017] at Sturminster Marshall proved 2.2 m of sandy grit and gravel. Boreholes farther west proved thicknesses exceeding 2.5 m, with the deepest [SY 9481 9988] proving more than 6.6 m of sand and gravel. The sequence there consists of 1 m of made ground, on 1 m of dark brown, silty clay with well-graded subangular flint gravel (topsoil), resting on over 6.6 m of well-graded, subangular flint gravel in a matrix of slightly sandy pale brown clay

Boreholes at the western end of the Wimborne Bypass proved up to 6.3 m of terrace deposits. The deposits are variable; the log of one borehole [SY 9994 9915] is reproduced below:

	Thickness m	*Depth* m
Topsoil, gravelly	0.3	0.3
3rd River Terrace Deposits		
Sand, clayey, very silty, with abundant gravel and carbonaceoous material (possible fill)	1.2	1.5
Clay, very silty, sandy, with abundant gravel	0.7	2.2
Sand, clayey, silty, with abundant gravel	0.5	2.7
Gravel, dark brown, loose, in a matrix of silty sandy clay	1.2	3.9
Gravel of rounded and subangular flints in a sand matrix	2.4	6.3
West Park Farm Member	6.2	12.5

Near Merley Hall Farm, the thickness of the terrace deposits varies from 4.3 m [SZ 0041 9921] to 6.3 m [SZ 0055 9915]. At both localities, an upper unit of clayey silty sand and gravel, 1.4 m and 1.8 m thick respectively, overlies gravel with sand.

Gravel, more than 2.5 m thick and set in a coarse-grained sand matrix, underlies the old railway embankment [around SZ 019 997] at Wimborne. Farther east, a borehole [SZ 0227 9958] proved 0.4 m of topsoil on 0.4 m of stiff clay with some gravel, on 1.3 m of gravel with some cobbles set in a sandy matrix. CRB

A borehole east of a tributary of Uddens Water [SU 0795 0303] proved 3 m of gravelly sand. Boreholes east of West Moors encountered up to 3.6 m of sand and gravel; others [SZ 1141 9902; 1202 9978] south of the Moors River revealed up to 0.6 m of topsoil and brickearth above 5.1 to 4.7 m of fine and coarse, subangular to subrounded flint gravel, with traces of quartz and sandstone in a matrix of coarse-grained sand. BJW

Second River Terrace Deposits

A borehole [SY 9715 9847] at Barford Dairy proved more than 3 m of gravel. On the opposite side of the river, a borehole [SY 9704 9849] at the waterworks encountered sand and gravel to a depth of 7.9 m. Farther east, a borehole [SY 9743 9851] proved 1.5 m of made ground, on 4.8 m of clayey, very gravelly sand, becoming a sandy gravel with depth.

Boreholes for the Wimborne Bypass proved thicknesses of sand and gravel ranging from 1.8 m [SZ 0304 9941] to 4.15 m [SZ 0270 9924]. In the latter, 1.8 m of clayey, sandy gravel overlies 2.35 m of rounded gravel in a sand matrix. Farther east, and south of the bypass, a borehole [SZ 0400 9908] encountered 3.3 m of angular and subangular flint gravel in a medium- to coarse-grained sand matrix; most pebbles are less than 6 cm across. CRB

At Hampreston [SZ 0501 9860], 2.8 m of gravel were proved. At Longham [SZ 0609 9730], 0.44 m of topsoil on 6.9 m of gravel were encountered. Near West Parley, a borehole [SZ 0858 9716] proved 3 m of sand and gravel.

South of Uddens Water, a borehole [SU 0740 0196] proved 0.5 m of topsoil on 4.5 m of sand and gravel; boreholes downstream show similar thicknesses. BJW

First River Terrace Deposits

South of Sturminster Marshall, the undulating surface of the deposits [SY 952 989; 954 988; 956 994; 958 991] is between 0.5 and 1 m above the floodplain. Up to 1.2 m of peat and peaty clay floor the hollows. A section [SY 9549 9932] south-east of the sewage farm showed 1 m of peat and peaty clay, resting on 0.6 m of grey, fine-grained sand with scattered flints, which in turn rests on more than 1 m of clayey sand and gravel. CRB

In the valley of Uddens Water, a borehole [SU 0804 0184] proved 0.5 m of peat on 3.5 m of gravelly sand, resting on sand of the Poole Formation. BJW

At Wick, the thickness of the sand and gravel (partially covered with made ground) varies between 4.3 m [SZ 1606 9170] and 6.7 m SZ 1599 9188], with base levels of 4.5 and 6.4 m below OD respectively. CRB

TERRACES OF THE RIVER AVON

Eighth River Terrace Deposits

In the north, near Sandford, disused gravel pits [SU 1736 0171] showed the following section; gravelly sand, 0.6 m; sandy, clayey, cryoturbated, yellow gravel, 2 m; fine- to coarse-grained, poorly sorted, clayey sand, 1 m; sandy gravel, over 0.2 m.

A borehole [SU 1746 0229] nearby proved:

	Thickness m	*Depth* m
8th River Terrace Deposits		
Gravel, sandy, coarse, with subangular flints	5.4	5.4
Clay, silty, yellowish orange	3.1	8.5
Gravel, sandy, with subangular flints	2.5	11.0
Barton Clay	2.0	13.0

Deposits of the Eighth Terrace, up to about 4.5 m thick, have been worked at several localities in the Avon Valley. One borehole [SZ 1905 9685] proved: topsoil, 0.5 m; fine and coarse, subangular to subrounded gravel with some sand, and containing scattered subrounded flint cobbles in basal 0.9 m, 3.9 m; on Barton Clay. A second borehole [SZ 1952 9627] proved: peaty dark brown sand, 0.3 m; fine and coarse, subangular gravel, with sand, 4.3 m; on Barton Clay. ECF

Gravel was formerly worked in a large 3 m-deep pit [SZ 198 946] west of Hinton Admiral Station. In the Burton Common No. 2 Borehole [SZ 1958 9495], 2.4 m of sand and gravel, with pebbles 2 to 3 cm in diameter and a sand layer between 1.8 and 2 m depths, overlie 1 m of sand, which in turn rests on 2 m of sandy gravel with pebbles up to 5 cm diameter. At Highcliffe [SZ 1980 9287], the gravel is up to 1.5 m thick and has a strongly cryoturbated base. CRB

Seventh River Terrace Deposits

Deposits of this terrace probably do not exceed 1.5 m in thickness. A small area of sandy gravel occurs east of Hurn Forest [SZ 127 996]. Near Hurn Forest [SZ 120 995], an impersistent veneer of cryoturbated gravel up to 1 m thick occurs.

Sixth River Terrace Deposits

Boreholes show thicknesses ranging from 4 to 9.4 m (Clarke, 1981), with the thickest sequence as follows:

	Thickness m	*Depth* m
6th River Terrace Deposits		
Soil and sandy silt, with fine-grained subangular flints	6.2	6.2
Gravel, sandy, with fine to coarse subangular to subrounded flint and a few subrounded sandstone clasts	3.2	9.4
Barton Clay	3.8	13.2

Farther south [SU 1633 0100], 2.8 m of loam were proved above 2.8 m of gravel. Boreholes nearer the western margin of the terrace penetrated up to 6.7 m of mainly sandy gravel. ECF

Up to 3 m of sand and gravel occur on Hengistbury Head; the base is locally strongly cryoturbated. A section [SZ 1729 9056] at the western end shows 3 m of sand and gravel, including a 1 m-thick bed of laminated sand in the middle. CRB

Fifth River Terrace Deposits

A borehole [SZ 1550 9964] in the northern part of the district, west of North Ripley, proved 1 m of sandy, silty clay ('brickearth'), on 2.1 m of fine and coarse, subangular to subrounded gravel with medium- to coarse-grained sand, on Branksome Sand. Another borehole [SZ 1637 9644], near Sopley Park, showed 1 m of fine-grained, sandy clay overlying 6 m of sandy, fine with coarse, subangular and subrounded flint gravel. A third borehole [SZ 1763 9851] proved 0.7 m of silty clay over 7.8 m of sandy gravel. Further south [SZ 1823 9565], near South Backhampton, 0.6 m of sandy loam overlies 5.1 m of fine to coarse, subangular gravel in a matrix of medium- to coarse-grained sand. ECF

Excavations for new houses [SZ 1892 9384] near Somerford proved at least 7 m of gravel. In the west, boreholes [SZ 1679 9433; 1666 9479] at Burton revealed between 2.6 and 5.3 m of sandy, cobbly gravel above Boscombe Sand.

Fourth River Terrace Deposits

The only extensive spread of Terrace 4 occurs south-south-east of St Catherine's Hill, where there are several 2 to 3 m-deep pits, most degraded or built over. In one pit [SZ 152 940] north of the railway line, however, up to 2.5 m of medium flint gravel is exposed along part of the north-eastern face. CRB

Third River Terrace Deposits

The surface of these deposits is rather hummocky, possibly because of the presence of small patches of blown sand. A borehole [SZ 1401 9674] north-west of Dudmoor Farm proved: clayey sand ('brickearth'), 1 m; on fine and coarse, subangular to subrounded flint gravel in a matrix of medium-grained sand, 3.4 m; on Branksome Sand, 7.4 m.

Near Burton, the thickness of the deposit exceeds 3 m. Grading figures show a sand:gravel ratio of 45:55 [SZ 1637 9399].

Second River Terrace Deposits

Near Week Farm, boreholes [SU 1340 0011; 1439 0048] proved 6.3 m and 8.2 m respectively of sandy gravel. Downstream, the deposits form three small areas of gravel within the floodplain. At the most northerly of these [SZ 1380 9914], 1.2 m of gravelly, brownish grey sand, overlie 0.2 m of fine gravel. A borehole at

Week Common [SZ 1343 9953] proved: topsoil, 0.2 m; very clayey sand ('brickearth'), 0.7 m; fine and coarse, subangular to subrounded flint gravel with a matrix of medium-grained sand, 5.9 m; on Branksome Sand. A borehole at Pithouse Farm [SZ 1372 9836] proved: topsoil, 0.2 m; coarse and fine subangular to subrounded flint gravel in a matrix of medium-grained sand, 6.4 m; on Branksome Sand. BJW

Boreholes near Purewell [around SZ 169 930] proved between 5.8 and more than 6.3 m of gravel. A borehole [SZ 1666 9285] 300 m south-west of the above, encountered 5.2 m of compact gravel.

CRB

First River Terrace Deposits

The surface of these deposits is just above floodplain level. North of Purewell, they exceed 3.6 m [SZ 1653 9374] in thickness. Grading figures show that they are dominantly medium gravels, with the sand content varying from 0 to 55 per cent [SZ 1655 9372 and 1653 9374]. On Stanpit Marshes, thicknesses of 10.5 [SZ 1715 9235], 8.6 [SZ 1687 9212] and 5.4 m [SZ 1958 9495] of terrace deposits have been proved. The Hengistbury No. 3 Borehole [SZ 1805 9072] proved about 17.5 m of sand and gravel above Branksome Sand; precise thicknesses of the various constituent lithologies were not determined because of poor sample recovery. CRB

ALLUVIUM

The alluvium consists of an upper unit of mottled dark grey and orange, commonly organic silt, silty clay and clayey sand, resting on a lower unit of sand and gravel. The thickness of the two units varies quite rapidly from 0.3 to 3.9 m for the upper, and from 0.35 to 6 m for the lower.

Locally, the alluvium of the Stour valley contains a rich molluscan fauna (Table 3). The assemblages are dominantly aquatic with minor terrestrial elements; they indicate hard, slow-flowing water with temperatures similar to those of the present day.

In the lower reaches of the rivers, the alluvium merges imperceptibly into estuarine alluvium. CRB

Details

River Stour and tributaries

Near Sturminster Marshall, thicknesses exceeding 1.5 m of greyish brown, shelly clay occur [ST 9540 0075], but in other nearby localities, the alluvial clay is 0.6 to 1 m thick [around ST 954 003]. A borehole [SY 9647 9967] near Barford Dairy proved 2.5 m of pale brown, sandy and silty clay, on 2.5 m of very coarse gravel, resting on broken chalk. Exposures along the river banks mostly show 1.4 m of mottled grey-brown, shelly clay. The fauna from two localities [SY 9659 9929; 9873 9936] near Barford Dairy is included in Table 3. Near the dairy, the upper clay unit thins to 0.4 m [SY 9595 9964].

Near Wimborne Minster [around SZ 005 995], the alluvium consists of 0.6 to 1.5 m of greyish brown, locally shelly clay above at least 1.4 m of gravel. Downstream [around SZ 033 992], shelly alluvial clay varies from 0.3 to more than 1.2 m in thickness and rests on gravel. Continuing downstream, an exposure [SZ 0435 9990] shows 1 m of greyish brown, shelly clay, with a concentration of shells at the base, resting on gravel. Mr D K Graham reports that the dominantly fluviatile molluscan assemblage (Table 3) also contains three terrestrial shells. Although the gastropod *Ovatella* usually occurs in estuarine situations, its occurrence in the fauna is confined to a single, doubtfully identified juvenile and there is otherwise no evidence of a tidal influence. CRB

Further downstream, a borehole [SZ 0774 9760] at Dudsbury penetrated 0.1 m of topsoil, on 1.9 m of silty clay, on 2.2 m of sandy gravel, on London Clay. At Ensbury, a borehole [SZ 0871 9663] proved 0.7 m of topsoil, on 4.2 m of silty clay, on sand of the Poole Formation.

In the northern part of the district, the alluvium in several boreholes along Uddens Water consists of sand and clay up to 2.8m thick. Downstream, boreholes [e.g. SU 0909 0152] reveal 3 m of clay, on 1 m of clayey peat, on 1 m of gravelly sand.

Near Hurn, the thickness varies from 0.5 [SZ 1427 9941] to 1.5 m [SZ 1170 9637]. Farther south, near Iford Bridge [SZ 1338 9430], 2 m of clayey coarse-grained sand with flint grains, overlie more than 1 m of sandy clay and clayey sand with gravel layers. BJW

River Avon

The only borehole [SU 1372 0170] in the northern part of the alluval tract proved 1.3 m of pebbly sand. An auger hole in a tributary stream bed [SU 1796 0229] penetrated 0.5 m of peaty brown clay on 0.75 m of clean sand, and a borehole [SU 1550 0062] in another tributary valley revealed 1.3 m of orange-brown and yellowish grey sandy clay, overlying gravel. South-west of Sopley, a borehole [SZ 1529 9657] showed 0.2 m of peaty loam, overlying 0.4 m of soft, sandy, silty, mottled orange and grey clay, resting on gravel. ECF

North of Christchurch, an upper unit of clayey peat at the top of the alluvium varies from 0.3 m to more than 1.2 m thick; the thickness of the alluvial gravel was proved at only one place [SZ 1561 9360] where it was 4.4 m thick. Boreholes [between SZ 1583 9303 and 1593 9290] on the east side of Christchurch proved, beneath made ground and ?head, 1.3 to 3 m of peaty alluvial clay, resting on 2 to 3.5 m of alluvial gravel. Towards Purewell, the thickness of black silty clay above gravel varies from 0.5 to 0.9 m [around SZ 163 927]. South of Christchurch, thicknesses in excess of 1.2 m of black peaty clay were consistently augered.

River Piddle system

In the western part of the district, the alluvium consists of at least 1.1 m of organic sandy clay [around SY 944 922].

A borehole [SZ 0191 9459] near Fleet's Corner proved 0.6 m of pale grey and orange-brown silty clay, above 1.8 m of medium-grained sand and gravel; the gravel content increases with depth.

The deposits are variable in the lower reaches of the Furze Brook. One auger hole [SY 9395 8564] proved 1.3 m of peaty clay; another [SY 9400 8702] passed through 0.6 m of peaty clay into gravel; a third [SY 9385 8552] proved 1.6 m of greyish brown clay. CRB

ESTUARINE ALLUVIUM

Estuarine Alluvium forms large marshy tracts mainly on the southern and western parts of Poole Harbour; the spreads appear to be growing. Between the Goathorn and Studland peninsulas, the 1:10 000 Ordnance Survey maps (1983) show areas of intertidal mud below High Water Mark, but these are now vegetated and lie above High Water Mark and are only flooded at exceptionally high tides. Considerable areas of older estuarine alluvium probably overlie tidal flat deposits. The deposit consists dominantly of an upper unit of mud and silt with a high organic content, overlying a sand and gravel unit; pockets of peat occur locally in the upper unit. ECF

Table 3 Mollusca from the Alluvium of the River Stour

	[SY 9659 9929]	[SY 9873 9936]	[SZ 0435 9990]
Gastropoda			
Acroloxus lacustris (Linné)	—	—	3
Aegopinella nitidula (Draparnaud)	—	—	6
Ancyclus fluviatilis (Müller)	38	1	26
Anisus vortex (Linné)	5	44	3
Armiger crista (Linné)	4	2	16
Bathyomphalus contortus (Linné)	43	23	·6
Bithynia tentacula (Linné)	617	2	59
B. tentacula [opercula]	132	2	12
B. leachii? (Sheppard)	—	—	3
B. sp. (juveniles)	—	—	90
Carychium minimum Müller*	—	5	1
Cochlicopa lubrica (Müller)	1	—	—
Discus rotundatus (Müller)*	1	—	1
Gyraulus albus (Müller)	124	4	46
Lymnaea palustris (Müller)	—	3	—
L. peregra (Müller) *ovata* Draparnaud	52	28	5
L. truncatula (Müller)	4	134	3
Lymnaea sp.	—	1	—
*Macrogastra rolphii?**	—	—	1
Oxyloma sp.	1	—	—
Phytia myosotis? (Draparnaud)	—	—	1
Planorbis planorbis (Linné)	2	—	1
Punctum pygmaeum? (Draparnaud)	—	4	—
Succinea oblonga (Draparnaud)	10	—	—
S. putris (Linné)	—	—	11
S. sp.	20	2	4
Theodoxus fluviatilis (Linné)	215	—	58
Trichia hispida (Linné)	28	11	—
Vallonia excentrica Sterki	—	9	—
Valvata cristata Müller	—	3	10
V. macrostoma Mörch	—	4	—
V. piscinalis (Müller)	355	—	101
gastropod indet.	—	fragments	—
Bivalvia			
Anodonta anatina (Linné)	1	—	—
Pisidium amnicum (Müller)	8	—	4
P. henslowanum (Sheppard)	—	—	5
P. milium Held	—	—	2
P. nitidum Jenyns	188	—	95
P. subtruncatum Malm	1	—	—
P. sp.	—	5	—
Sphaerium corneum (Linné)	23	—	5
S. corneum? (fragments)	18	—	—

* terrestrial species

The composition of the lower gravelly deposits varies from a gravelly sand to a sandy gravel. The sand fraction, which may form up to 95 per cent, is dominantly medium grained; the gravel fraction is dominantly fine to medium. The clasts are mainly angular and subangular flints (81 to 90 per cent), with between 1 and 8 per cent well-rounded flints, 3 to 8 per cent quartz and 1 to 4 per cent ironstone (Mathers, 1982).

CRB

Details

A borehole [SY 9815 9184] at Turlin Moor showed, beneath 1.55 m of made ground, 1.1 m of soft black organic clay, on 0.95 m of light grey sand and fine gravel. A borehole south of Poole Harbour [SY 9489 8671] proved 0.5 m of brown clay, above 0.5 m of peaty clay, on 2.8 m of gravel. Another borehole nearby [SY 9399 8722] showed 1.5 m of peat, on 0.8 m of dark grey silty clay, overlying 2.7 m of silty, grey-brown peat, above 5 m of sand and gravel.

Estuarine Alluvium on the north-eastern side of Hole's Bay is covered by made ground. Boreholes near Fleet's Corner [around SZ 010 932] prove a dominantly arenaceous alluvial succession above the Poole Formation. For example, one borehole [SZ 0097 9308] penetrated 0.3 m of soft silty topsoil, above 2.7 m of sandy gravel. Another [SZ 0115 9333] passed through 0.9 m of sandy clay with peat and gravel, into 2.1 m of grey-brown sand with a trace of shells. A third [SZ 0101 9329], however, proved 2 m of made ground on 0.8 m of soft, dark grey, very silty clay that rested on 1.1 m of soft peat and dark grey silty, peaty clay. At Stanley Green, up to 1.5 m of sand overlies 1 m of sandy gravel [SZ 0101 9289; 0102 9279]. East of Fleets Lane, 0.3 to 1 m of peat was encountered at depths varying from 0.9 to 1.5 m from the surface.

At Sterte, a borehole [SZ 0114 9208] proved: made ground, 1.8 m; soft dark grey, very sandy clayey peat, 0.5 m; soft brown-black peat, 0.5 m; loose, very silty, gravelly medium- and coarse-grained sand, 1.2 m. Upstream, the following section [SZ 0160 9207] was seen: topsoil and made ground, 0.4 m; on gravel 0 to 0.15 m; on sand, medium-grained, with scattered flints, 0.8 m; on sand and gravel, 0.4 m; on sand, not penetrated. A nearby borehole [SZ 0166 9215] proved 0.25 m of topsoil on 1 m of mottled, yellowish grey, fine-grained, organic sand with gravel, on 0.8 m of black, laminated, slightly clayey peat, on Oakdale Clay.

A borehole [SZ 0078 9176] near Poole Power Station proved, beneath made ground, 0.4 m of grey silt resting on 1.5 m of sand and gravel in a clay matrix. At Lower Hamworthy, the estuarine alluvium incorporates a shell bank. Boreholes [around SZ 008 902] at Ferry Road showed, beneath 0.6 to 4 m of made ground, up to 3.3 m of 'oyster' shells, resting on a variable sequence of shelly soft clay, silty clay, clayey silt and silty sand to a depth of at least 14 m. Boreholes south of the railway [around SZ 0067 9007] proved a thinner drift sequence, with gravel up to 5 m thick, resting on sand of the Poole Formation. The most easterly borehole proved 8.5 m of estuarine alluvium as follows:

	Thickness m	*Depth* m
Made ground	2.8	2.8
Estuarine Alluvium		
Gravel, compact, with stones up to 50 mm	1.5	4.3
Clay, silty, soft, with some sand and small stones	0.7	5.0
Clay, silty, dark grey, firm, becoming softer with depth	3.3	8.3
Silt, soft, grey, with peat	1.0	9.3
Sand, medium-grained, dark, with some stones up to 50 mm	2.0	11.3
Poole Formation	7.3	18.6

At the gasworks [SZ 018 905], between 1 and 3.3 m of soft, dark brown-grey, silty clay has been proved beneath made ground, and resting on the Poole Formation. ECF, CRB

PEAT

Much of the floodplain of the River Allen is occupied by peat, 0.2 to more than 1.3 m thick, overlying gravel. A small spread of peat, about 0.75 m thick, occurs on the north side of the Moors River alluvium [SZ 117 985]. In the south, Hartland Moor is a large boggy tract of peat of unknown thickness. CRB

STORM GRAVEL BEACH DEPOSITS

The most extensive storm-gravel spit extends north-eastwards from Hengistbury Head, and almost closes the mouth of Christchurch Harbour. Another spit, about 400 m long, now concealed beneath made ground, existed in the area known as The Baiter [SZ 021 902] on the west side of Parkstone Bay (Freshney et al., 1985). Minor gravel spits, up to 500 m long, but generally not more than 1 m high, occur around Christchurch Harbour (Freshney et al., 1984).

Burton (1931) documented the changes in the extent of the Christchurch Harbour spit and the former positions of The Run [SZ 185 918]. At its maximum, in 1880, the spit extended almost to Cliff End [c. SZ 198 927], approximately 1.8 km north-east of its position in 1847, and 1 km beyond its present position. CRB

BLOWN SAND

Fine-grained, well-sorted sand covers an area of about 1.5 km^2 on Studland Peninsula, where it forms east-north-east-trending dunes up to 6 m high. The dunes lie in three major tracts which form a series of accreting ridges, the oldest being to the north-west and the latest against the sea on the south-east (Diver, 1933). Diver showed that the ridges of dunes were built up against the Redhorn Peninsula, composed mainly of solid outcrops, in the 17th century. A tidal estuary or lagoon that was forming by 1721 in the lee of the earliest dune tract became increasingly enclosed during the late 18th century and, by the end of the century, became cut off as a lagoon into which only Spring tides flowed. The lagoon was finally completely isolated by the most easterly dune belt, and today forms a body of water known as the Little Sea. Because these dune sands form a prograding sequence, they almost certainly overlie older beach sands of unknown thickness as far west as the solid outcrop near the road to the ferry [SZ 030 857]. ECF

Other smaller areas of blown sand occur on cliff tops, and consist of well-sorted, fine-grained sand up to 4 m thick, blown up the cliff face and deposited in an area of slackening wind velocity just behind the cliff tops. It is still accumulating where the cliff top is not built on [e.g. SZ 1220 9135]. Blown sand overlying the shingle spit that extends north-north-east from Hengistbury Head probably has an origin similar to that seen on top of the cliffs. Burton (1931) recorded that the dunes on the spit had been up to 5 m high, but that their height was much reduced by 'trampling feet'. ECF, CRB

Blown sand occurs on the surface of some of the terrace deposits in the Avon valley (Freshney et al., 1984), where it consists of up to 3 m of well-sorted, fine-grained sand probably derived from Tertiary sands on St Catherine's Hill. BJW

LANDSLIP

Landslips occur at several localities near Lytchett Matravers. All are developed on clays of the West Park Farm Member and are associated with springs issuing from the base of the overlying Warmwell Farm Sand. The most spectacular slip [SY 945 977], 650 m long by 200 m across, occurs west of Combe Almer. The slips have taken place on slopes of 10° or less. Slopes in the mixed sand and clay sequences of the higher part of the London Clay or Poole Formation are locally as steep as 25° and appear to be stable. CRB

SINK HOLE OR SOLUTION-COLLAPSE HOLLOWS

Swallow holes were recognised by Reid in 1894 during the original geological survey of the north-west part of the district. In places, the individual holes were recorded; elsewhere the general note 'swallows' was written on the map. White (1917) noted that the holes 'lie mostly in the paths of intermittent brooks which collect the surface-drainage of the adjacent slopes on the Eocene beds; for swallow holes, in the early stage of their development, thrive best on light and occasional draughts of acidulated soil-water and are liable to be choked by sediment when the inflow is rapid or sustained'.

It can be seen from Figure 27 that the majority of the depressions are developed on the sand members of the London Clay. The sandy strata no doubt facilitate the passage of ground water to the Chalk. The bigger depressions occur in the larger dry or intermittently dry valleys. In the district, their maximum diameter is about 50 m and their depth is up to 2 to 3 m; 60 or 70 m diameter depressions up to 4 m deep have been noted farther west (Bristow, 1987c). In places, two holes may partially overlap to give a dumb-bell shaped depression.

In addition to the depressions along the major valleys, some of which have recognisable swallow holes in their floors, there are many saucer-shaped depressions on the valley sides (Figure 27), varying from 5 to 40 m across and up to 2 m deep. Whilst one or two on Tertiary strata away from the valley bottoms may be old pits, their large number, concentrated in areas where the Chalk lies at shallow depth, suggests that they are collapse structures developed over solution cavities in the Chalk. Such hollows are locally aligned at right angles to valley sides [e.g. around SY 9612 9627 and SY 9495 9611] or follow the courses, commonly straight [e.g. around SY 9538 9650], of minor dry valleys. This suggests that some of the collapses may be joint controlled. Mr Guest, of Sunnyside Dairy, who has filled in several of these depressions, states that 'they continue to grow over the years'. CRB

MADE GROUND

Extensive areas of marshland, low-lying ground bordering rivers, several small valleys and many pits have been reclaimed in recent years. Much of the fill, which may be up to 10 m thick, is domestic waste, but industrial refuse, spoil from sand and brick pits, and imported sand and dredged material from Poole Harbour also form a significant part of the made ground. The known areas of made ground and their presumed compositions are shown in Freshney et al. (1984, fig.18; 1985, fig.22) and Bristow and Freshney (1986a, fig.13). CRB, ECF

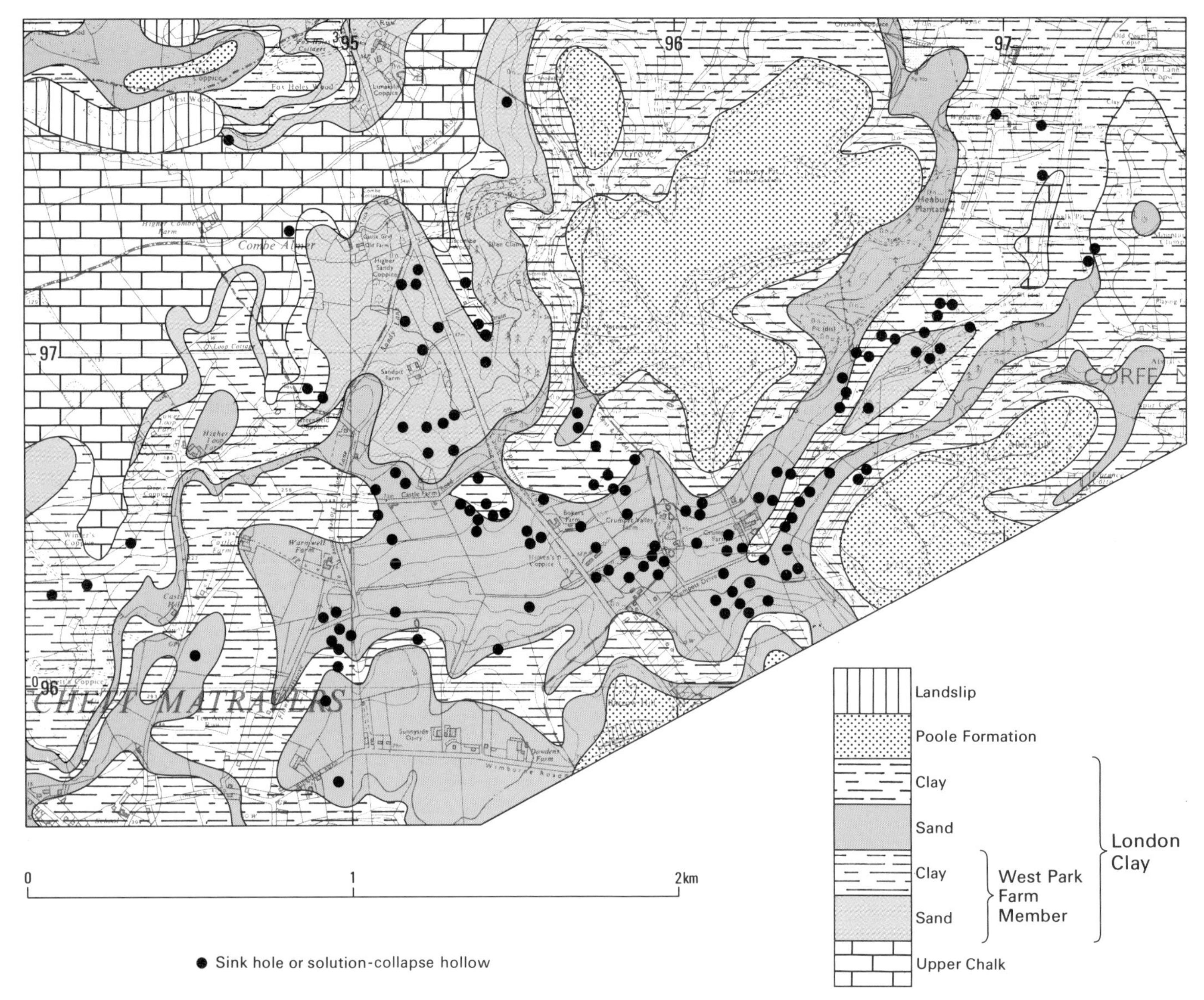

Figure 27 Distribution of sink holes or solution-collapse hollows in the Corfe Mullen–Lytchett Matravers area.

TEN

Economic geology

SAND AND GRAVEL

West of the River Avon, most of the sand and gravel terrace deposits have been sterilised by urban development, except for a small part of Canford Heath and small areas around Merley House [SZ 010 981] and Knighton [SZ 042 974], where the only working sand and gravel pit is located (Table 4). East of the Avon, extensive Eighth River Terrace Deposits occur in the Bransgore to Hinton Admiral area, where they have been widely exploited in the past.

The lower terraces and alluvial gravel, extensively developed along much of the Stour valley, are largely sterilised by urban development. Potential resources are present near Parley (estimated at about 70 million m^3 by Clarke (1981)), between Longham and Wimborne, and around Sturminster Marshall. There are insufficient data to calculate the resources in these last two areas, but the Second River Terrace gravels are at least 5 m thick; the alluvial gravel is up to 6 m thick.

The terraces of the River Avon form extensive flats and, except in its lower reaches where they are extensively built over, form a potential resource of over 180 million m^3 (Clarke, 1981). The higher terraces east of the Avon are estimated to have resources of 75 million m^3. These have been exploited, but the only working pit [SZ 215 011] is a small-scale Forestry Commission one on Holmsley Ridge (Table 4). CRB, ECF

SAND

The Beacon Hill Brick Co. Ltd makes sand/lime bricks using sand of the Poole Formation mixed with lime which is brought to the site (Table 4). Production in 1985 was 43 322 tons.

Sand of the Poole Formation is also dug from the Henbury Pit [SY 964 975] and used for building, horticulture and brickmaking (Table 4). Some sand is sold 'as dug' without being screened; other sand is screened for building purposes. Plastering sand, which is also produced at this pit, is a mixture of washed and screened sand.

On Canford Heath [SZ 031 967], sand of the Poole Formation is worked (Table 4). Between 300 000 and 500 000 tonnes of building, mortar, asphalt, plastering and concreting sand, together with single-size aggregate, are produced annually. CRB

The Boscombe Sand, which has an extensive outcrop, is too fine-grained for use in the building industry.

The Chama Sand and Becton Sand were once worked for moulding sand and glass sand respectively, in areas east of the district. ECF

BRICK AND POTTERY CLAYS

Six stratigraphical units have been worked for brick and pottery clays. These are the reddened clays of the West Park Farm Member of the London Clay, the younger grey clays of the same formation, and clays in the Poole, Branksome Sand, Barton Clay and Headon formations.

The Knoll Manor Pit of Pilkington & Carter [SY 974 978] is the only pit in which clays of the West Park Farm Member are currently worked (Table 4); it produces 2500 to 3000 tons of unglazed floor tiles per year. Some of the clay from this pit is transported to a pottery at Stoke on Trent. Clays of the West Park Farm Member were also formerly worked in a

Table 4 Working pits in the district

Name	NGR	Formation/Member	Applications
Knoll Manor	SY 974 978	London Clay (West Park Farm Member)	Floor tiles
Canford Heath	SZ 031 967	Poole Formation	Building, mortar, asphalt, plastering and concreting sand
Henbury	SY 964 975	Poole Formation	Building, plastering and horticultural sand, together with sand for brickmaking
Beacon Hill	SY 980 950	Poole Formation	Sand/lime bricks
Arne	SY 975 896	Poole Formation (Oakdale Clay)	Wall and floor tiles, and refractory products
Knighton	SZ 042 974	River Terrace Deposits	Hoggin and concreting gravel
Holmsley Ridge	SU 215 001	River Terrace Deposits	Hoggin for tracks

pit [SY 989 987] at Candy's Lane. Working of this latter pit changed from opencast to adit mining in about 1950. The adits, about 30 m long, radiated out from the face. However, this method of working proved uneconomic and the pit closed in the late 1950s.

London Clay has been worked for brickmaking in many small pits in the Lytchett Matravers area and near Holt in the north, but all have long been abandoned. CRB

Two main types of clay occur in the Poole Formation: one is brown, carbonaceous, laminated and very sandy, and the other more homogeneous, grey and commonly red-stained, with a lower sand content. The brown sandy clays were used mainly for brickmaking, and the grey clays for pottery. The Oakdale and Parkstone clays were the main source of pottery clay in the present district (Freshney et al., 1985).

Gilkes (1978) noted an eastward diminution in the kaolinite content in Tertiary clays from the Wareham ballclay area, towards the eastern Hampshire Basin. In the Poole–Bournemouth area, the kaolinite content is possibly too low to produce the best white-firing ballclays.

Details

Details of many of the former pits and the uses to which the clay was put can be found in Freshney et al., (1984; 1985), Bristow and Freshney (1986a) and Young (1972). A few notes on the larger, more recent workings are included below.

The Creekmoor Clay was dug for brick clay in pits south-west of Waterloo [SZ 005 936]; these pits are now filled and are being built over.

The Oakdale Clay has been extensively exploited in the west of the district. One of the largest pits was worked by the former Hamworthy Junction Brickworks (later part of the Kinson Pottery Company) [SY 988 915], which manufactured bricks and glazed pipeware until about 1965. Clay from the pits at Lake [SY 982 908] was worked until about 1950 for the manufacture of pipes, bricks and insulators; much of the clay was transported to Stoke on Trent. Until 1978, a seam of grey ball clay in the Oakdale Clay was dug from a pit [SY 951 921] south-west of Lytchett Minster. This clay was blended with clays from south Devon and used for frost-resistant tiles. At Arne [SY 975 896], the ball clay is dug and blended with other Dorset clays and used principally in the production of wall and floor tiles and refractory products (Table 4).

At present, the only working brick pit [SY 98 95] is that of the Beacon Hill Brick Co. Ltd. Previously, bricks were made at this site using the local Broadstone Clay, but now sand/lime bricks are made from sand of the Poole Formation mixed with lime which is brought to the site. Some sand from the Henbury Pit (see below) is also used for brickmaking.

Broadstone Clay appears to have been worked only for brick clays, mainly around Beacon Hill, Broadstone, and north-eastwards towards Canford Magna and near Wimborne.

The Parkstone Clay was the main source of pottery clay in the present district; the major pits were mainly in the Foxholes [SZ 030 930] to Mannings Heath [SZ 038 945] area on the south side of Canford Heath. On Brownsea Island, the Parkstone Clay was also used extensively for bricks and pipes until 1887. ECF, CRB

Clays of the Branksome Sand were dug for brickmaking in the King's Park and Queen's Park areas of Bournemouth, as well as in other small pits in the Bournemouth to Parley area (Freshney et al., 1985). The brickclays of the Branksome Sand consist mainly of laminated, commonly carbonaceous, sandy clay and silty clayey sand often described as 'loams'. Because of the lenticular nature of the clays, the pits tended to be small, and because of the extensive gravel cover, they were mostly located at the margins of gravel spreads.

Clays of the Headon Formation and the Barton Clay have been worked for bricks near Bransgore (Freshney et al., 1984). ECF

BUILDING STONE

White (1917) noted that ferruginously cemented sandstones of the West Park Farm Member and Lytchett Matravers Sand of the London Clay in the Lytchett Matravers area, have been used in local buildings. He quoted examples of their use in the walls of Wimborne Minster and in the bridge at Sturminster Marshall. CRB

LIME AND MARL

Lime and marl were dug from many chalk pits in the north-western part of the district during the 19th century. One pit [ST 958 007] was worked well into the 20th century. CRB

HYDROCARBONS

The Wytch Farm Oilfield (Table 5; frontispiece) was discovered in December 1973 when oil-bearing Bridport Sands, which produced oil at an average rate of 600 barrels per day on drill stem test in Wytch Farm No. 1 Borehole (Colter and Havard, 1981; Hinde, 1980), were proved. Subsequent exploration proved a larger, deeper, Sherwood Sandstone reservoir which flowed at a rate of 3300 barrels per day on drill stem test in December 1977. With 'reserves' of 230 × 10^6 barrels (Anon., 1984), it is the biggest onshore oilfield in north-west Europe.

The oil-bearing structures lie within an east–west-trending, northerly dipping, pre-Cretaceous tilt-block. Closure to the south at both reservoir levels is formed by the southerly dipping, normal southern boundary fault (Figure 28). Dip closure confines the eastern and western flanks of both reservoir levels, as it does to the north of the upper reservoir, though some minor normal faults are also important. Closure on the northern flank of the lower reservoir, however, is formed by a northerly dipping, normal boundary fault. Mudstones of the Mercia Mudstone Group and the Fuller's Earth cap the Sherwood Sandstone and Bridport Sands reservoirs respectively. Each also forms the downfaulted seals of their subjacent reservoir formations.

Free oil is contained in the shelly levels of the Oxford Clay, and good oil-shows usually occur in fractures and vugs in the Cornbrash and Forest Marble. A sporadically developed sandstone in the latter has flowed oil on drill stem test.

The Bridport Sands reservoir ranges in porosity from less than 10% in the hard, well-cemented beds, to 32% in the intervening friable sandstones. Permeability ranges correspondingly from 0 to 300 millidarcies. The primary porosity of the Sherwood Sandstone reservoir is rejuvenated; it has been preserved by early formed anhydrite or calcite cement, which was subsequently dissolved away (Penn et al., 1987). The upper part of the sandstone succession contains 70% of

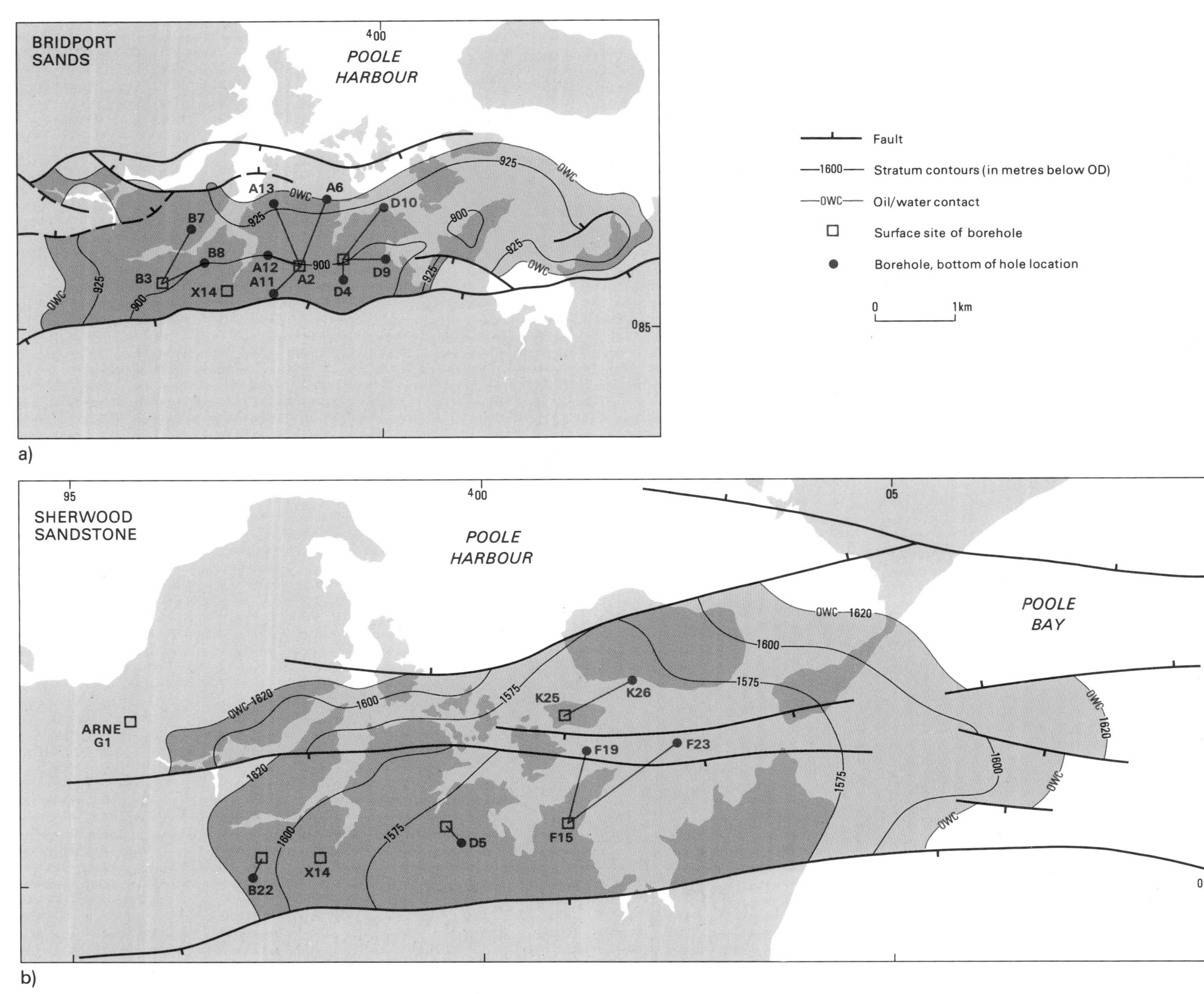

Figure 28 Wytch Farm Oilfield: a) depth to top of the Bridport Sands reservoir (based on Hinde, 1980, fig. 2), b) depth to top of the Sherwood Sandstone reservoir, (based on Hinde, 1980, fig. 9; Dranfield et al., 1987, fig. 1).

Table 5 Wytch Farm Oilfield: data summary based on Anon. (1984), Hinde (1980), Colter and Havard (1981) and Dranfield et al. (1987)

Formation	Bridport Sands	Sherwood Sandstone
Age	Toarcian	?Anisian – Ladinian
Trap type	Tilted fault-block	Tilted fault-block
Depth to crest (m)	- 887	- 1588
Depth to oil/water contact (m)	- 936	- 1615
Oil column (m)	10 – 32	4 – 29 (sandstones)
Permeability (millidarcies)	0 – 300	0.01 – 7000 (sandstones)
Temperature (°C)	42	61
Pressure (pounds per sq. in. at datum)	1492 (at - 923 m)	2442 (at - 1584 m)
Oil gravity (API)	35 – 37°	38 – 42°
Gas/oil ratio (atmos. standard cu.ft/barrels)	152	365
Production: oil (barrels per day)	2,500	60,000 (forecast)
gas (million cubic feet per day)	10	
Oil reserves, total field (barrels)		230×10^6

the oil in place. Its sandstones range in porosity from 4% to 29%, and in permeability from 0.01 to 7000 millidarcies (Colter and Havard, 1981). The porosity and permeability of the Sherwood Sandstone, however, are both considerably reduced by detrital clay present as discrete lenses in, and in the pore spaces of, the sandstones (Table 6). The effective porosity and permeability (Table 7) has been modelled by Dranfield et al. (1987).

Throughout the Wessex Basin, mudstones of the Lias, Oxford Clay and Kimmeridge Clay are rich hydrocarbon source rocks, yielding Kerogen Types II, III and a mixture of II and III, with a potential for generating oil and gas (Ebukanson and Kinghorn, 1985). They show considerable range in maturity, however, depending on their structural situations (Penn et al., 1987). They are probably of rich potential throughout the district, but in the Wytch Farm area they are immature, since their structural position, on the South Dorset High, prevented their burial at sufficient depths (and therefore temperatures) over a long enough period to generate hydrocarbons in quantity (Colter and Havard, 1981; Ebukanson and Kinghorn, 1986).

Table 6 Wytch Farm Oilfield: porosity and permeability of Sherwood Sandstone reservoir rock types. Based on Dranfield et al. (1987, fig.4)

Lithology	Average porosity %	Mean permeability (millidarcies)
Sandstone	19.6	489
Muddy sandstone	14.7	14
Muddy siltstone	11.6	0.35

Table 7 Wytch Farm Oilfield: effective porosity and permeability of Sherwood Sandstone stratigraphical subdivisions. Based on Dranfield et al. (1987, fig.4)

Layer	Effective porosity (%)	Effective permeability, horizontal (millidarcies)	Effective permeability, vertical (millidarcies)
Upper Sandstone	13.00 – 14.8	30.1 – 36.3	0.025 – 0.44
Middle Sandstone	16.9	49.3	0.348
Lower Sandstone	15.0	20.0	0.095

The origin of the oil in the reservoirs is therefore thought to be from outside the district, probably from the Lower Lias in the Channel Basin to the south. There, mid-Mesozoic down-faulting most probably ensured sufficient depth of burial for hydrocarbons to be generated during the Cretaceous period (Colter and Havard, 1981; Penn et al., 1987). The complex, northward migration path that this explanation demands has been modelled by Selley and Stoneley (1987). They suggest that successive extensional and compressive phases of tectonism could have led to the current entrapment at levels structurally higher than the putative source-rock material, by alternate passage and sealing along fault-planes.

IEP

HYDROGEOLOGY AND WATER SUPPLY

The district lies across the boundary between Hydrometric Areas 43 and 44. The water resources are managed by the Wessex Water Authority. The district lies mainly within Unit 4 of the Wessex Water Authority, with a small part in the south-west being within Unit 5 (Monkhouse and Richards, 1982). Information on the area is also published in the Hydrological Survey for the Wessex Rivers (Anon., 1967), and further details are given by White (1917).

The central part of the district rises in places to more than 90 m above sea level. The northern flank of the higher ground drains to the River Stour which flows approximately east-south-eastwards to be joined by the River Avon near Christchurch. The southern flank is drained by small streams flowing directly into the sea in Lytchett Bay, Poole Harbour and Poole Bay.

The mean annual rainfall is about 900 mm, and the mean annual evaporation probably about 450 mm. Infiltration into the Chalk outcrop, which in this district is largely covered by drift, is probably about 300 mm/yr, while that into the outcrop of the Palaeogene sands is probably not less than 350 mm/yr.

Most of the water supply is taken from sources outside the district. However, surface water is taken from intakes on the River Stour, both for public and for private (industrial) supply. In the west, some 33 megalitres per day (Mld) are available from boreholes in the Chalk at Corfe Mullen [SY 974 983], while a further 20 Mld can be taken from a site near Sturminster Marshall, also from the Chalk. Four borehole sources are licensed for industrial use in the central tract, and there are shallow wells supplying groundwater for agricultural and domestic use.

Chalk, the major aquifer in the Wessex Water Authority area, crops out in the valley of the River Stour, although there it is almost completely concealed by drift. The junction with the overlying Palaeogene strata dips gently south and south-eastwards until, at the coast, it lies at a depth of between 130 m in the west and 330 m and more in the east.

In the Stour valley, and beneath a thin cover of London Clay, the Chalk has a mean yield of 11.5 litres per second (l/sec) for a drawdown of 10 m from boreholes of 300 mm diameter penetrating 30 m into the saturated aquifer.

Near Wimborne Minster and northwards, boreholes of 200 mm diameter, penetrating 50 m of saturated Chalk beneath London Clay, would be expected on average to yield some 3.5 l/sec. As the thickness of cover increases, yields tend to become less, and near the coast, a borehole of similar dimensions might be expected to yield less than 0.6 l/sec for a drawdown of 10 m.

A deep borehole drilled for water at Christchurch [SZ 1545 9380] in 1905, penetrated Chalk from 251 m to 281 m depth, but little water was obtained. Interestingly, an unusually high yield of more than 30 l/sec was taken from seams of lignite in the Branksome Sand at a depth of some 60 m, but the water was so rich in iron that the borehole had to be closed off, and was subsequently abandoned.

The quality of Chalk groundwater in the north is generally good, with a total hardness of between 250 and 400 mg/l and a chloride ion concentration of about 30 mg/1. Towards the coast, information is limited, but the chloride ion concentration appears to increase to more than 200 mg/l, and possibly more than 500 mg/1.

Chalk becomes less permeable under increasing cover, and this, combined with the necessarily deeper (and therefore more costly) boreholes required to tap the aquifer, makes the development of borehole sources unattractive.

Moderate yields of groundwater have been recorded from the sandy beds (where present) of the West Park Farm Member of the *London Clay*. However, most of this member consists of clay, and where sandy horizons are developed within this formation, principally in the west, they contain a significant clay and silt fraction, permeability is usually low, and the West Park Farm Member is generally regarded as an aquiclude.

The higher part of the London Clay is for the most part an aquiclude. Small supplies of groundwater can be obtained from the thicker sandy beds present in the west, and from the weathered zone, but even these yield water only reluctantly to wells and boreholes. The quality is usually poor, and lack of proper well design can cause pollution from surface drainage. The thicker sands, such as the Warmwell Farm Sand, do throw out springs of some size, even as much as 50 m^3/d. However, the water quality is usually poor with high concentrations of iron and sulphate.

Potentially, the *Poole Formation* forms a good aquifer consisting of sandy beds separated by clays. Nevertheless, relatively little use has been made of this aquifer other than a few shallow wells for domestic supplies. Boreholes with properly designed filter packs and sand screens should be capable of yielding up to 3.5 l/sec, possibly up to 8.0 l/sec.

The groundwater quality in the Poole Formation should generally be good. The total dissolved solids content should not exceed 300 mg/1 or the total hardness 200 mg/1. The chloride ion concentration should be less than 30 mg/1, although adjacent to the coast, higher concentrations might be expected. Nitrate and sulphate concentrations should also be low. However, iron may often be present in concentrations of more than 1.0 mg/1.

The Branksome Sand and Boscombe Sand have similar hydrogeological characteristics and may be considered as a single multilayered aquifer. These strata contain the only significant groundwater resources of the district. Groundwater levels are commonly close to the ground surface, and boreholes not infrequently overflow.

Allowing for the run-off from urban areas, the average annual replenishment to the Palaeogene sands is probably about 10 to 15 million cubic metres per annum. Because of the multilayered nature of the aquifer, it would in practice be difficult to develop this potential to the full.

Boreholes of 200 mm diameter penetrating 15 m of saturated sand would be expected to have an average yield of about 6.0 l/sec for a drawdown of 5 m. For a diameter of 300 mm, the average yield should approach 8.0 l/sec for the same drawdown. Sand screens with properly designed filter packs (Monkhouse, 1974) would be necessary, both to support the borehole wall and to prevent the ingress of sand; natural filter packs could probably be developed in the coarser sands. Boreholes penetrating running sand have often failed, or even collapsed, where inadequately designed and developed sand screens have been fitted.

A study of well yields in the Branksome Sand and Boscombe Sand and their lateral equivalents over the whole of the Hampshire Basin, has shown that yields in the Bournemouth–Christchurch area seem to be rather less than elsewhere in the basin. Table 8 shows the yields that might be expected in a given percentage of cases from these formations in the district. It is assumed that 30 m of screen are present in the saturated aquifer, and that the drawdown is 5 m.

In general within the Hampshire Basin, borehole yields do not appear to increase significantly at depths greater than about 60 m beneath the potentiometric surface. There is insufficient information in the present district with which to construct contours of the potentiometric surface. In general, levels have not been recorded more than 10 m above OD;

Table 8 Expectancy of water yields (in l/secs) from the Branksome Sand and Boscombe Sand

Borehole diameter (mm)	Percentage of cases		
	75%	50%	25%
150	3.7	6.4	11.3
300	5.1	8.9	15.9
450	6.2	10.9	19.3

they are near sea level close to the coast, and rise gently inland.

The groundwater quality is usually fairly good with total dissolved solids mostly less than 250 mg/l. The total hardness (as $CaCO_3$) is generally less than 100 mg/l and is not known to exceed 200 mg/l, while the chloride-ion concentration is normally less than 40 mg/l. Iron, however, is often present in concentrations of greater than 1.0 mg/l, and may be sufficiently high to cause water-treatment problems.

In areas adjacent to the coast, prolonged pumping may lead to seawater intrusion, and the siting of boreholes less than 500 m from the shore is not recommended.

The *Barton Clay* forms an aquiclude and does not yield useful supplies of groundwater. The *Chama Sand,* the *Becton Sand* and the *Headon Formation* are present only in the east of the district. Small supplies (probably less than 0.5 l/sec) might be obtained from the Chama and Becton sands, but the grain-size tends to be very fine, and it would be difficult to install satisfactory sand screens in boreholes.

Drift deposits are generally too thin and of too limited an extent to comprise a useful groundwater resource. Where permeable drift overlies the Palaeogene sands, the two are usually in hydraulic continuity, and it is customary to construct boreholes into the solid formation rather than into the superficial.

Supplies greater than 10 l/sec have been obtained in the east from the thicker *river terrace deposits,* but only where these extend below Ordnance Datum.

A number of shallow wells and boreholes have in the past taken groundwater from river terrace deposits in the Stour valley. One site north of Corfe Mullen yielded up to 390 m^3/d for public supply. Terraces at higher levels tend to drain rapidly and thus have low storage capacities. Boreholes are not generally employed to abstract groundwater from the river terrace deposits and large (up to 3 m) diameter shafts or tube-wells are preferred. The groundwater quality can be good, but with the water table close to the ground surface, the supplies are vulnerable to pollution from surface sources. Where abstraction sites are located near water courses, there is a possibility of induced recharge and the groundwater quality may then reflect the surface-water quality. Most of the sources in these superficial deposits are no longer in use.

The *alluvium* of the district generally has a high clay content and yields little water. Its hydrogeological significance lies in its ability to limit the infiltration of river water into the underlying deposits.

Protection of aquifers

The Control of Pollution Act 1974 and the relevant EEC Council Directive (Anon., 1979) require that groundwater be protected from pollution. The vulnerability of an aquifer (as distinct from a source such as a well or borehole) may be regarded as a measure of the ease with which pollutants may pass from ground surface to the saturated zone. Considering the multilayered nature of the Palaeogene aquifer, it is difficult to propose any definite rules for its protection on this basis. In general, pollutants may quickly reach the saturated zone in sandy aquifers where the water table is close to ground surface. In a multilayered succession, the less permeable layers will inhibit the spread of pollution so that only a part of the aquifer may be endangered in a specific case.

Protection of sources

In aquifers where the water table is close to ground surface, overground drainage can reach the saturated zone quickly, the more so if the well or borehole is poorly constructed. It is inadvisable to locate sources close to polluting agencies such as farm yards, stables or waste disposal sites. Drainage from roads, particularly major roads, can also be a hazard where it is discharged to soakaways.

Excavated shafts should have a watertight coping and be surrounded by a concrete apron. Boreholes should be fitted with a length, or lengths, of plain lining tube extending from the ground surface at least to the saturated zone, and this tube should be sealed externally with cement grout. RAM

Underground workings

Clay has been mined in underground workings at Hamworthy, Beacon Hill, Corfe Mullen and Brownsea Island. Unfortunately, there is little detail about the type and extent of the workings. White (1917, p.70) noted that clay was mined at Hamworthy [= Ham] Common. This is almost certainly a reference to adits which were driven northwards in Oakdale Clay from the pit at Lake. Reid shows the position of one such mine [c.SY 9808 9095] on his manuscript geological map (Bristow and Freshney, 1986, p.84). The Beacon Hill Pottery mined white clay of the Broadstone Clay in the northern part of their pit [SY 9760 9462]. The pit and the adits were abandoned some time prior to 1894 (Bristow and Freshney, 1986, p.85). The working of the pit at Candy's Lane, Corfe Mullen, is described on p.98. On Brownsea Island, Parkstone Clay was extracted from below sea level in underground shafts. The position of several of these infilled shafts can still be seen [around SZ 0170 8854 and 0202 8845] (Bristow and Freshney, 1987, p.37).

CRB, EFC

REFERENCES

Most of the references listed below are held in the Library of the British Geological Survey at Keyworth, Nottingham. Copies of the references can be purchased subject to the current copyright legislation

ALLEN, D J, and HOLLOWAY, S. 1984. *The Wessex Basin. Investigation of the geothermal potential of the UK.* 80pp. (Keyworth, Nottingham: British Geological Survey.)

ANON. 1967. *Wessex Rivers Hydrological Survey. Hydrometric Areas Numbers 42, 43 and 44.* Ministry of Housing and Local Government. (London: HMSO.)

— 1979. The protection of groundwater against pollution caused by certain dangerous substances. Council Directive of 17th December, Official Journal of the European Communities, L20/4348.

— 1984. New pipeline to boost oil flow from Wytch Farm field. *Oil and Gas Journal,* Vol. 85, 24.

BAILEY, H W, GALE, A S, MORTIMORE, R N, SWIECICKI, A, and WOOD, C J. 1983. The Coniacian–Maastrichtian Stages of the United Kingdom, with particular reference to southern England. *Newsletter of Stratigraphy,* Vol. 12, 29–42.

BLONDEAU, A, and POMEROL, C. 1969. A contribution to the sedimentological study of the Palaeogene of England. *Proceedings of the Geologists' Association,* Vol. 79, 441–455 [for 1968].

BRISTOW, C R. 1987a. *Geological notes and local details for 1:10 000 sheets ST 90 SW and SE (Shapwick–Pamphill, Dorset).* (Exeter: British Geological Survey.)

— 1987b. *Geological notes and local details for 1:10 000 sheet SU 00 SW (Wimborne, Dorset).* (Exeter: British Geological Survey.)

— 1987c. *Geological notes and local details for 1:10 000 sheets SY 99 NW and SW (Morden and Morden Heath, Dorset).* (Exeter: British Geological Survey.)

— and FRESHNEY, E C. 1986a. *Geology of Sheets SY 99 NE and SE and parts of SY 99 NW and SW. Corfe Mullen–Lytchett Minster, Dorset. Geological report for DOE: Land Use Planning.* (Exeter: British Geological Survey.)

— — 1986b. *Geology of the Poole–Bournemouth area. Geological report for DOE: Land Use Planning.* (Exeter: British Geological Survey.)

— — 1987. *Geological notes and local details for 1:10 000 sheets SY 98 NW, NE, SW and SE and parts of SZ 08 NW and NE (Arne–Wytch Farm area, Dorset).* (Exeter: British Geological Survey.)

BUJAK, J P, DOWNIE, C, EATON, G L, and WILLIAMS, G L. 1980. Dinoflagellate cysts and acritarchs from the Eocene of southern England. *Special Papers in Palaeontology, Palaeontological Association of London,* No. 24.

BURTON, E ST JOHN. 1931. Periodic changes in position of The Run at Mudeford near Christchurch, Hants. *Proceedings of the Geologists' Association,* Vol.42, 157–174.

— 1933. Faunal horizons of the Barton Beds in Hampshire. *Proceedings of the Geologists' Association,* Vol. 44, 131–167.

BURY, H. 1933. The Plateau Gravels of the Bournemouth area. *Proceedings of the Geologists' Association,* Vol. 44, 314–35.

BUURMAN, P. 1980. Palaeosols in the Reading Beds (Paleocene) of Alum Bay, Isle of Wight, UK. *Sedimentology,* Vol. 27, 593–606.

CALKIN, J B, and GREEN, J F N. 1949. Palaeoliths and terraces near Bournemouth. *Proceedings of the Prehistorical Society,* Vol. 15, 21–37.

CHADWICK, R A. 1985. Permian, Mesozoic and Cenozoic structural evolution of England and Wales in relation to the principles of extension and inversion tectonics. 9–25 in *Atlas of onshore sedimentary basins in England and Wales: Post-Carboniferous tectonics and stratigraphy.* WHITTAKER, A. (editor). (Blackie, Glasgow.)

— 1986. Extension tectonics in the Wessex Basin. *Journal of the Geological Society of London,* Vol. 143, 465–488.

CHANDLER, M E J. 1960. Plant remains of the Hengistbury and Barton Beds. *Bulletin of the British Museum (Natural History), Geology,* Vol. 4, 191–238.

— 1962. *The Lower Tertiary floras of southern England.* II. Flora of the Pipe-clay Series of Dorset (Lower Bagshot). (London: British Museum (Natural History).)

— 1963. *The Lower Tertiary floras of southern England.* III. Flora of the Bournemouth Beds; the Boscombe, and the Highcliff Sands. (London: British Museum (Natural History).)

CHAPMAN, F. 1913. On some foraminifera from the Eocene Beds of Hengistbury Head, Hampshire. *Geological Magazine,* Vol. 10, 555–559.

CLARKE, M R. 1981. The sand and gravel resources of the country north of Bournemouth, Dorset. Description of parts of 1:25 000 sheets SU 00, 10, 20, SZ 09, 19, and 29. *Mineral Assessment Report Institute of Geological Sciences,* No. 51.

CLASBY, P S. 1971. Report of field meeting to the Barton on Sea district of Hampshire. *Tertiary Times,* Vol. 1, 115–116.

— 1972. Report of field meeting to Barton on Sea, Hampshire. *Tertiary Times,* Vol. 2, 51–52.

COLTER, V S. and HAVARD, D S. 1981. The Wytch Farm Oil Field, Dorset. In *Petroleum geology of the continental shelf of north-west Europe.* ILLING, L V, and HOBSON, G D (editors). (London: The Institute of Petroleum.)

COOPER, J, HOOKER, J J, and WARD, D J. 1976. Report of field meeting to east Dorset (including Holt Wood and Studland Bay). *Tertiary Research,* Vol. 1, 3–4.

COPE, J C W, GETTY, T A, HOWARTH, M K, MORTON, N, and TORRENS, H S. 1980. A correlation of Jurassic rocks in the British Isles. *Special Report of the Geological Society of London,* No. 14, 1–109.

COSTA, L I. and DOWNIE, C. 1976. The distribution of the dinoflagellate *Wetzeliella* in the Palaeogene of north-western Europe. *Palaeontology,* Vol. 19, 591–614.

— DOWNIE, C, and EATON, G L. 1976. Palynostratigraphy of some Middle Eocene sections from the Hampshire Basin, England. *Proceedings of the Geologists' Association,* Vol. 87, 273–284.

COX, B M, 1982. A note on the biostratigraphy of the Kimmeridge Clay cored interval of the Winterborne Kingston borehole, Dorset. 45–46 in *The Winterborne Kingston borehole, Dorset, England.* RHYS, G H, LOTT, G K, and CALVER, M A (editors). *Report of the Institute of Geological Sciences,* No. 81/3.

— and GALLOIS, R W. 1981. The stratigraphy of the Kimmeridge Clay of the Dorset type area and its correlation with some other Kimmeridge sequences. *Report of the Institute of Geological Sciences,* No 80/4, 1–44.

Curry, D. 1942. The Eocene succession at Afton brickyard, IOW. *Proceedings of the Geologists' Association,* Vol. 53, 88–101.

— 1976. The age of the Hengistbury Beds (Eocene) and its significance for the structure of the area around Christchurch, Dorset. *Proceedings of the Geologists' Association,* Vol. 87, 401–407.

— Adams, C G, Boulter, M C, Dilley, F C, Eames, F E, Funnell, B M, and Wells, M K. 1978. A correlation of Tertiary rocks in the British Isles. *Geological Society of London Special Report,* No. 12.

Daley, B, and Edwards, N. 1971. Palaeogene warping in the Isle of Wight. *Geological Magazine,* Vol. 108, 399–405.

Daniels, M C. 1970. Report of Easter field meeting to Barton, Hants, and surrounding regions. *Tertiary Times,* Vol. 1, 27–28.

Diver, C. 1933. The physiography of South Haven Peninsula, Studland Heath, Dorset. *Geographical Journal,* Vol. 81, p.404.

Dranfield, P, Begg, S M, and Carter, R R, 1987. Wytch Farm Oilfield: reservoir characterization of the Triassic Sherwood Sandstone for input to reservoir simulation studies. 149–160 in *Petroleum geology of north west Europe.* Brooks, J, and Glennie, K (editors). (London: Graham and Trotman).

Eaton, G L. 1976. Dinoflagellate cysts from the Isle of Wight, southern England. *Bulletin of the British Museum (Natural History), Geology,* Vol. 26, 225–332.

Ebukanson, E J, and Kinghorn, R R F, 1985. Kerogen facies in the major Jurassic mudrock formations of southern England and their implication on the depositional environments of their precursors. *Journal of Petroleum Geologists,* Vol. 8, 435–62.

— — 1986. Maturity of organic matter in the Jurassic of southern England and its relation to the burial history of the sediments. *Journal Petroleum Geologists,* Vol. 9, 259–280.

Edwards, R A. and Freshney, E C. 1987a. Lithostratigraphical classification of the Hampshire Basin Palaeogene Deposits (Reading Formation to Headon Formation). *Tertiary Research,* Vol. 8, 43–73.

— — 1987b. The geology of the country around Southampton. *Memoir of the Geological Survey of Great Britain, Sheet 315 (England and Wales).*

Fisher, O. 1862. On the Bracklesham Beds of the Isle of Wight Basin. *Quarterly Journal of the Geological Society of London,* Vol. 18, 65–94.

Freshney, E C, and Bristow, C R. 1987. *Geological notes and local details for 1:10 000 sheets SU 10 SE, 20 SW, SZ 29 NW and 29 SW (Ringwood–Barton on Sea).* (Exeter: British Geological Survey.)

— — and Williams, B J. 1984. *Geology of sheet SZ 19 (Hurn–Christchurch, Dorset and Hants). Geological report for DOE: Land Use Planning.* (Exeter: British Geological Survey.)

— — — 1985. *Geology of Sheet SZ 09 (Bournemouth–Poole–Wimborne, Dorset). Geological report for DOE: Land Use Planning.* (Exeter: British Geological Survey.)

Gale, A S, Wood, C J, and Bromley, R G. 1987. The lithostratigraphical and marker bed correlation of the White Chalk (Late Cenomanian to Campanian) in southern England. *Mesozoic Research,* Vol. 1, 107–118.

Gardner, J S. 1879. Description and correlation of the Bournemouth Beds. Part 1. Upper Marine Series. *Quarterly Journal of the Geological Society of London,* Vol. 35, 209–228.

— 1882. Description and correlation of the Bournemouth Beds. Part II. Lower or Freshwater Series. *Quarterly Journal of the Geological Society of London,* Vol. 38, 1–15.

Gatrall, M, Jenkyns, H C, and Parsons, C F. 1972. Limonitic concretions from the European Jurassic with particular reference to the 'snuff boxes' of southern England. *Sedimentology,* Vol. 18, 79–103.

Gilkes, R J. 1978. On the clay mineralogy of Upper Eocene and Oligocene sediments in the Hampshire Basin. *Proceedings of the Geologists' Association,* Vol. 89, 43–56.

Green, J F N. 1946. The terraces of Bournemouth, Hants. *Proceedings of the Geologists' Association,* Vol. 57, 82–101.

— 1947. Some gravels and gravel-pits in Hampshire and Dorset. *Proceedings of the Geologists' Association,* Vol. 58, 128–143.

Henson, M R. 1970. The Triassic rocks of south Devon. *Proceedings of the Ussher Society,* Vol. 2, 172–177.

— 1972. The form of the Permo-Triassic basin in south-east Devon. *Proceedings of the Ussher Society,* Vol 2, 447–457.

Highley, D. 1975. Ball clay. *Mineral Dossier Mineral Resources Consultative Committee,* No. 11.

Hinde, P. 1980. The development of the Wytch Farm Oilfield. *Communication Institute of Gas Engineers, London,* No. 1133, 1–19.

Hooker, J J. 1975. Report of a field meeting to Hengistbury Head and adjacent areas, Dorset, with an account of published work and some new exposures. *Tertiary Times,* Vol. 2, 109–121.

— 1986. Mammals from the Bartonian (middle/late Eocene) of the Hampshire Basin, southern England. *Bulletin of the British Museum (Natural History) Geology,* Vol. 39, 191–478.

Holloway, S, Milodowski, E E, Strong, G E, and Warrington, G. 1989. The Sherwood Sandstone Group (Triassic) of the Wessex Basin, southern England. *Proceedings of the Geologists' Association,* Vol. 100, 383–394.

Jukes-Browne, A J. 1904. Cretaceous rocks of Britain, 3. The Upper Chalk of England. *Memoir of the Geological Survey of Great Britain.*

Keen, D H. 1980. The environment of deposition of the south Hampshire Plateau Gravels. *Proceedings of the Hampshire Field Club Archaeological Society,* Vol. 36, 15–24.

Keeping, H. 1887. On the discovery of the *Nummulina elegans* Zone at Whitecliff Bay, Isle of Wight. *Geological Magazine,* Vol. 4, 70–72.

King, C. 1981. The stratigraphy of the London Clay and associated deposits. *Tertiary Research Special Paper,* No. 6.

Knox, R W O'B, Morton, A C, and Lott, G K. 1982. Petrology of the Bridport Sands in the Winterborne Kingston borehole, Dorset. 107–121 *in* The Winterborne Kingston borehole, Dorset, England. Rhys, G H, Lott, G K, and Calver, M A (editors). *Report of the Institute of Geological Sciences,* No. 81/3.

Lott, G K. 1982. The sedimentology of the Lower Chalk (Middle–Upper Cenomanian) of the Winterborne Kingston borehole, Dorset. 28-34 *in* The Winterborne Kingston borehole Dorset, England. Rhys, G H, Lott, G K. and Calver, M A (editors). *Report of the Institute of Geological Sciences,* No.81/3.

— Sobey, R B, Warrington, G. and Whittaker, A. 1982. The Mercia Mudstone Group (Triassic) in the western Wessex Basin. *Proceedings of the Ussher Society,* Vol. 5, 340–346.

Lott, G K, and Strong, G E. 1982. The petrology and petrography of the Sherwood Sandstone (?Middle Triassic) of the Winterborne Kingston borehole, Dorset. 135–142 *in* The Winterborne Kingston borehole, Dorset, England. Rhys, G H, Lott, G K. and Calver, M A (editors). *Report of the Institute of Geological Sciences,* No. 81/3.

LYELL, C. 1827. On the strata of the Plastic Clay Formation exhibited in the cliffs between Christchurch Head, Hampshire, and Studland Bay, Dorsetshire. *Transactions of the Geological Society of London,* Ser. 2, Vol. 2, 279–286.

MATHERS, S J. 1982. The sand and gravel resources of the country between Dorchester and Wareham, Dorset. Description of parts of 1:25 000 sheets SY 68, 69, 78, 79, 88, 89, 98 and 99. *Mineral Assessment Report Institute of Geological Sciences,* No. 103.

MELVILLE, R V, and FRESHNEY, E C. 1982. *British Regional Geology: the Hampshire Basin.* (London: HMSO for Institute of Geological Sciences.)

MILODOWSKI, A E, STRONG, G E, WILSON, G E, ALLEN, D J, HOLLOWAY, S, and BATH, A H. 1986. *Diagenetic influences on the aquifer properties of the Sherwood Sandstone in the Wessex Basin. Investigations of the geothermal potential UK.* 83pp. (Keyworth, Nottingham: British Geological Survey.)

MONKHOUSE, R A. 1974. *An assessment of the groundwater resources of the Lower Greensand in the Cambridge–Bedford (Region. Reading: Water Resources Board.)*

— *and* RICHARDS, H J. 1982. *Groundwater resources of the United Kingdom.* Commission of the European Communities. (Hannover: Th. Schufer.)

MORTER, A A. 1982. The macrofauna of the Lower Cretaceous rocks of the Winterborne Kingston borehole, Dorset. 35–38 *in The Winterborne Kingston borehole, Dorset, England.* RHYS, G H, LOTT, G K, and CALVER, M A (editors). *Report of the Institute of Geological Sciences,* No. 81/3.

MORTIMORE, R N. 1986. Stratigraphy of the Upper Cretaceous White Chalk of Sussex. *Proceedings of the Geologists' Association,* Vol. 97, 97–139.

MORTON, A C. 1982. Heavy minerals of Hampshire Basin Palaeogene strata. *Geological Magazine,* Vol. 119, 463–476.

ORD, W T. 1914. The geology of the Bournemouth to Boscombe Cliff section. *Proceedings of the Bournemouth Natural Science Society,* Vol. 5, 118–135.

PENN, I E. 1982. Middle Jurassic stratigraphy and correlation of the Winterborne Kingston borehole, Dorset. 53–76 in The Winterborne Kingston borehole, Dorset, England. RHYS, G H, LOTT, G K, and CALVER, M A (editors). *Report of the Institute of Geological Sciences,* No. 81/3.

— 1985. Some aspects of the deep geology of southern England and their bearing on the deep geology of France. *Programme Géologie Profonde de la France; Deuxième phase d'investigation, 1984/1985 documents BRGM,* Vol. 95/12, 1–28.

— CHADWICK, R A, HOLLOWAY, S, ROBERTS, G, PHARAOH, T C, ALLSOP, J M, HULBERT, A G, and BURNS, I M. 1987. Principal features of the hydrocarbon prospectivity of the Wessex-Channel Basin, UK. 109–118 in *Petroleum geology of north west Europe.* BROOKS, J, and GLENNIE, K (editors). (London: Graham and Trotman.)

— DINGWALL, R J, and KNOX, R W O'B. 1980. The Inferior Oolite (Bajocian) sequence from a borehole in Lyme Bay, Dorset. *Report of the Institute of Geological Sciences,* Vol. 79/3, 1–27.

— MERRIMAN, R J, and WYATT, R J. 1979. The Bathonian strata of the Bath–Frome area. *Report of the Institute of Geological Sciences,* Vol. 78/22, 1–88.

PLINT, A G. 1980. Sedimentary studies in the Middle Eocene of the Hampshire Basin. Unpublished DPhil. thesis, University of Oxford (3 volumes).

— 1982. Eocene sedimentation and tectonics in the Hampshire Basin. *Journal of the Geological Society of London,* Vol. 139, 249–254.

— 1983a. Sandy fluvial point-bar sediments from the Middle Eocene of Dorset, England. *Special Publication of the International Association of Sedimentologists,* No. 6, 355–368.

— 1983b. Facies, environments and sedimentary cycles in the Middle Eocene, Bracklesham Formation of the Hampshire Basin: evidence for global sea-level changes. *Sedimentology,* Vol. 30, 625–653.

— 1983c. Liquefaction, fluidization and erosional structures associated with bituminous sands of the Bracklesham Formation (Middle Eocene) of Dorset, England. *Sedimentology,* Vol. 30, 525–535.

— 1988. Sedimentology of the Eocene strata exposed between Poole Harbour and High Cliff, Dorset, UK. *Tertiary Research,* Vol. 10, 107–145.

PRESTWICH, J. 1846. On the Tertiary or Supracretaceous Formations of the Isle of Wight, etc. *Quarterly Journal of the Geological Society of London,* Vol. 2, 255–259.

— 1849. On the position and general characters of the strata exhibited in the coast section from Christchurch Harbour to Poole Harbour. *Quarterly Journal of the Geological Society of London,* Vol. 5, 43–49.

REED, F R C. 1913. Notes on the Eocene Beds of Hengistbury Head. *Geological Magazine,* Vol. 10, 101–103.

REID, C. 1898. The geology of the country around Bournemouth. *Memoir of the Geological Survey of Great Britain,* Sheet 329 (England and Wales).

— 1902. The geology of the country around Ringwood. *Memoir of the Geological Survey of Great Britain,* Sheet 314 (England and Wales).

RHYS, G H, LOTT, G K, and CALVER, M A (editors). 1982. The Winterborne Kingston borehole, Dorset, England. *Report of the Institute of Geological Sciences* 81/3, 196.

SEALY, K R. 1955. The terraces of the Salisbury Avon. *Journal of the Royal Geographical Society,* Vol. 121, 350–356.

SELLEY, R C, and STONELEY, R. 1987. Petroleum habitat in south Dorset 139–148 *in* BROOKS, J, and GLENNIE, K (editors). *Petroleum geology of north west Europe.* (London: Graham and Trotman.)

SMITH, D B, BRUNSTROM, R G W, MANNING, P I, SIMPSON, S. and SHOTTON, F W. 1974. A correlation of Permian rocks in the British Isles. *Special Report of the Geological Society of London,* No. 5, 1–45.

STINTON, F. 1975; 1977. Fish otoliths from the English Eocene. *Palaeontographical Society* [Monograph], Part 1, 1–56 (1975); Part 2, 57–126 (1977).

TYLOR, A. 1850. On the occurrence of productive iron ore in the Eocene formations of Hampshire. *Quarterly Journal of the Geological Society of London,* Vol. 6, 133–134.

WARRINGTON, G. 1974. Les évaporites du Trias britannique. *Bulletin de la Société Géologique de France,* Vol. 16, 708–723.

— AUDLEY-CHARLES, M G, ELLIOTT, R E, EVANS, W B, IVIMEY-COOK, H C, KENT, P E, ROBINSON, P L, SHOTTON, F W, and TAYLOR, F M. 1980. A correlation of Triassic in the British Isles. *Geological Society of London Special Report,* No. 13.

— and SCRIVENER, R C. 1988. Late Permian fossils from Devon: regional geological implications. *Proceedings of the Ussher Society,* Vol. 7, 95–96.

West, G H. 1886. *in* Report of the committee ... appointed for the purpose of inquiring into the rate of erosion of the sea coasts of England and Wales, and the influence of the artifical abstraction of shingle or other material in that action.6. Christchurch to Poole. Topley, W (editor). *Report of the British Association for the Advancement of Science for 1885,* 427–428.

White, H J O. 1917. Geology of the country around Bournemouth (2nd edition). *Memoir of the Geological Survey of Great Britain,* Sheet 329 (England and Wales).

Whitaker, W. 1910. The water supply of Hampshire. *Memoir of the Geological Survey of Great Britain.*

Whitaker, W. and Edwards, W. 1926. Wells and springs of Dorset. *Memoir of the Geological Survey of Great Britain.*

Whittaker, A (editor) 1985. *Atlas of onshore sedimentary basins in England and Wales.* 1–68. (Glasgow: Blackie.)

Whittaker, A, Holliday, D W, and Penn, I.E. 1985. Geophysical logs in British stratigraphy. *Geological Society of London Special Report,* No.18.

Williams, B J. 1987. *Geology of Sheets SU 00SE and SU 10SW (West Moors–St. Leonards, Dorset).* (Exeter: British Geological Survey.)

Wood, C J, Bigg, P J, and Medd, A.W. 1982. The biostratigraphy of the Upper Cretaceous (Chalk) of the Winterborne Kingston borehole, Dorset. 19–27 *in* The Winterborne Kingston borehole, Dorset, England. Rhys, G H, Lott, G K, and Calver, M A (editors). *Report of the Institute of Geological Sciences,* No. 81/3.

Wright, C A. 1972. The recognition of a planktonic foraminiferid datum in the London Clay of the Hampshire Basin. *Proceedings of the Geologists' Association,* Vol. 83, 413–420.

Wright, T. 1851. A stratigraphical account of the section at Hordwell, Beacon and Barton Cliffs on the coast of Hampshire. *Annals and Magazine of Natural History,* Vol. 7, 433–466.

Young, D. 1972. Brickmaking in Dorset. *Proceedings of the Dorset Natural History & Archaeological Society,* Vol. 93, 213–242.

APPENDIX

Geological Survey photographs

Copies of the photographs are deposited for reference in the British Geological Survey library at Keyworth, Nottingham NG12 5GG. Prints and slides may be purchased. The more recent photographs listed below were taken by Messrs M J Pulsford, H J Evans and B Starbuck, and are available in colour and black and white. They belong to Series A.

A 5810	Section in 'Bagshot Beds' [Branksome Sand], Parkstone
A 5811	Cliffs and chine in 'Bagshot Beds' [Branksome Sand], Durley Chine
A 5812	River cliff section in 'Bagshot Beds' [Branksome Sand], Durley Chine
A 5813	Cliff section in 'Bracklesham Beds', Boscombe
A 5814	Gravel-spit at the mouth of the Avon–Stour Estuary
A 5815	Hengistbury Head
A 5816	Sand dunes and bar at the mouth of the Avon–Stour Estuary
A 5817	Bar at the mouth of the Avon–Stour Estuary
A 5818	Old workings for iron ore, Hengistbury
A 5819	Estuary of the Avon–Stour
A 5820	Old iron ore workings, Hengistbury
A 5821	Cliff section in 'Bracklesham Beds' [Barton Group] Hengistbury
A 5822	Cliff section of gravel overlying 'Barton Beds'
A 5823	Cliff section of gravel overlying 'Barton Beds'
A 5824	Cliff section of gravel overlying 'Barton Beds'
A 12243	Bar at the mouth of the Avon–Stour Estuary.
A 12244	Bar at the mouth of the Avon–Stour Estuary.
A 12688	'Bracklesham Beds' [Poole Formation], St Catherine's Hill
A 12689	'Plateau Gravels' [Twelfth River Terrace Deposits], St Catherine's Hill
A 12690	Cryoturbated 'Plateau Gravels' [Thirteenth River Terrace Deposits], St Catherine's Hill
A 12708	'Valley gravels' of the River Stour, Hengistbury Head
A 12709	[Boscombe Sands and Barton Clay], Hengistbury Head
A 12710	'Upper Hengistbury Beds' [Boscombe Sand and Barton Clay], Hengistbury Head.
A 12711	'Plateau Gravels' [Thirteenth River Terrace Deposits], Holmsley Gravel Pit
A 12712	Cryoturbated 'Plateau Gravels' [Thirteenth River Terrace Deposits], Holmsley Gravel Pit
A 12723	'Ball clay' in 'Bagshot Beds' [Poole Formation], Canford Heath
A 12724	Trough cross-bedded sands [Poole Formation], Canford Heath
A 12725	Trough cross-bedded sands [Poole Formation], Canford Heath. (Close up)
A 12726	Contorted bedding in 'Bagshot Beds' [Poole Formation], Canford Heath.
A 12728	Cryoturbated 'Plateau Gravels' [Thirteenth River Terrace Deposits], Holmsley Ridge

FOSSIL INDEX

GENERAL INDEX

BRITISH GEOLOGICAL SURVEY

Keyworth, Nottingham NG12 5GG
(06077) 6111

Murchison House, West Mains Road,
Edinburgh EH9 3LA 031-667 1000

London Information Office, Natural History Museum
Earth Galleries, Exhibition Road, London SW7 2DE
071-589 4090

The full range of Survey publications is available through the Sales Desks at Keyworth and at Murchison House, Edinburgh, and in the BGS London Information Office in the Natural History Museum Earth Galleries. The adjacent bookshop stocks the more popular books for sale over the counter. Most BGS books and reports are listed in HMSO's Sectional List 45, and can be bought from HMSO and through HMSO agents and retailers. Maps are listed in the BGS Map Catalogue and the Ordnance Survey's Trade Catalogue, and can be bought from Ordnance Survey agents as well as from BGS.

The British Geological Survey carries out the geological survey of Great Britain and Northern Ireland (the latter as an agency service for the government of Northern Ireland), and of the surrounding continental shelf, as well as its basic research projects. It also undertakes programmes of British technical aid in geology in developing countries as arranged by the Overseas Development Administration.

The British Geological Survey is a component body of the Natural Environment Research Council.

Maps and diagrams in this book use topography based on Ordnance Survey mapping

HMSO publications are available from:

HMSO Publications Centre
(Mail and telephone orders)
PO Box 276, London SW8 5DT
Telephone orders 071-873 9090
General enquiries 071-873 0011
Queueing system in operation for both numbers

HMSO Bookshops
49 High Holborn, London WC1V 6HB
071-873 0011 (Counter service only)
258 Broad Street, Birmingham B1 2HE
021-643 3740
Southey House, 33 Wine Street, Bristol BS1 2BQ
(0272) 264306
9 Princess Street, Manchester M60 8AS
061-834 7201
80 Chichester Street, Belfast BT1 4JY
(0232) 238451
71 Lothian Road, Edinburgh EH3 9AZ
031-228 4181

HMSO's Accredited Agents
(see Yellow Pages)

And through good booksellers